THE FIRST WORD

THE FIRST WORD

The search for the origins of language

CHRISTINE KENNEALLY

Viking

VIKING

Published by the Penguin Group

Penguin Group (USA) Inc., 375 Hudson Street,

New York, New York 10014, U.S.A.

Penguin Group (Canada), 90 Eglinton Avenue East, Suite 700, Toronto,

Ontario, Canada M4P 2Y3 (a division of Pearson Penguin Canada Inc.)

Penguin Books Ltd, 80 Strand, London WC2R 0RL, England

Penguin Ireland, 25 St. Stephen's Green, Dublin 2, Ireland (a division of Penguin Books Ltd)

Penguin Books Australia Ltd, 250 Camberwell Road, Camberwell,

Victoria 3124, Australia (a division of Pearson Australia Group Pty Ltd)

Penguin Books India Pvt Ltd, 11 Community Centre,

Panchsheel Park, New Delhi–110 017, India

Penguin Group (NZ), 67 Apollo Drive, Rosedale, North Shore 0745, Auckland, New Zealand

(a division of Pearson New Zealand Ltd.)

Penguin Books (South Africa) (Pty) Ltd, 24 Sturdee Avenue,

Rosebank, Johannesburg 2196, South Africa

Penguin Books Ltd, Registered Offices: 80 Strand, London WC2R 0RL, England

First published in 2007 by Viking Penguin, a member of Penguin Group (USA) Inc.

5 7 9 10 8 6 4

LIBRARY OF CONGRESS CATALOGING-IN-PUBLICATION DATA

Kenneally, Christine.

The first word / by Christine Kenneally.

p.   cm.

Includes bibliographical references and index.

ISBN 978-0-670-03490-1

1. Language and languages. 2. Evolution. I. Title.

P107.K465   2007

400—dc22   2007003182

Printed in the United States of America

Set in Perpetua

Designed by Francesca Belanger

For Agnes (Nessie) Kenneally

CONTENTS

THE FIRST WORD

Prelude

Imagine all of your knowledge about language whirling above your head instead of inside it, each word a star. At least sixty thousand glittering specks of light— "brick," "axel," "pawn," "shoe," "Victorian," and "apple"—hang in the air over your skull.

Look more closely at each star and you'll see that each is not a single point of light but an intense cluster of all the things you know about that word. The "rose" star includes bits of knowledge such as how the word sounds when you say it and how it looks when you write it. Perhaps a small image of a rose floats there, or maybe ten or twenty rose prototypes, all of which help you connect "rose," the word, to the bloom of any Rosaceae shrub you come across.

You know that roses, like most flowers, are perfumed, delicate, and short-lived. This constitutes physical knowledge. When you raise one to your nose, your body has expectations about what's going to happen next. If the flower smells rotten, you feel shock as well as distaste. You also have linguistic knowledge. "Rose" has a special re-lationship with words like "scent" and "fragrance"; they go together in a way that "concrete" and "fragrance," for example, do not.

In the word constellation now twisting above your head, picture the connection between "rose" and "scent" as a filament running between the two stars. Other lines would run between "rose" and "red," and "rose" and "flower," and "rose" and "nose." In fact, lines would connect "rose" and all sorts of words—words with similar mean-ings, words that make similar sounds, words that are the same part of speech.

If all the different things you know about "rose" and its connections with other words were embodied in your language universe, lines would rapidly proliferate.

Try mapping out connections for other English words, including everything from the lousiest pun to the densest nugget of grammar, all the associations and the con-jugations, the synonyms, homonyms, and homophones—make them manifest.

Now everywhere you look, fibers wind around words, tugging them together,

pulling the entire assemblage tight. Some links might have especially significant relationships. The stronger the connection between words, the thicker the thread will
be. (Consider "mow" and "lawn," for example.) There are so many lines you can
hardly see the words for the relationships between them. What began with a few
threads is now a tangled language web.

It may seem as if the complicated mass above you maps the world. After all, the
connections between words are like the connections between physical objects. People
eat apples, for example, and not coincidentally there is a word for people, a word
for eating, and a word for apple. Actors act on objects in language as they do in life,
and when we put words together, it seems as if the point of language is simply and
accurately to describe the real thing. But language is not a replication of the physical world. If you look closely, you'll see there are holes in the web you have built,
places where the world of words does not correspond to the physical world. Words
align according to their own rules. Moreover, there is so much that happens in our
daily lives for which there is no language. There is no verb for the way an airplane's
shadow ripples across the landscape, no specific adjective to describe that single un-
ruly hair poking out of your eyebrow.

Because language does not mimic the world, you can do things with it that are
impossible under the laws of physics. You are a god in language. You can create. Destroy. Rearrange. Shove words around however you like. You can make up stories
about things that never happened to people who never existed. You can push a camel
through the eye of a needle. It's easy if "camel" and "needle" are words.

In language, mortality does not tick relentlessly. You can conceive of yourself as
alive forever. Or you can imagine yourself dead. And then alive again. You can live,
die, live, die, live, die, live.

Now imagine what it's like for the person nearest you. Look over and visualize
everything he knows about language bristling above his head. No individual's language is ever exactly the same as another's, but assuming he also speaks English, the
basic size and shape of his word constellation are similar to yours. Perhaps he has
stars that you don't, like "Rexism," an early-twentieth-century religious movement.
Maybe the connections that run from his "who" and his "whom" are different from
yours. Because dialects and idioms differ, some words and grammatical rules aren't
identical, even if you both speak English.

But there is much that is the same between your language and his. Your "rose"
star maps closely onto his, as do "car" and "grill" and "Bombay" and tens of thou-

sands of other words. If threads ran from the words in your language network to the matching words in his, your heads would be simultaneously joined and dwarfed by the vast, complicated lattice you share.

If you add a third and fourth person, it's going to start to look as if your heads are plugged into a huge, shining network consisting of billions of lines. Add some more speakers, and the number of connections becomes uncountable.

After you've included everyone who speaks English, imagine the 410 million people in the world who speak Spanish. Draw a different network for each of them individually, and then join it up. Bring in Kurdish and Arabic. Add Basque, Urdu, and American Sign Language.

Now add all the speakers of all the languages of the world. Step back. You can see that the whirling word universe above your own head is merely an atom in the linguistic matrix surrounding your entire species. Everyone who has language is connected, and anyone who is connected lives in two worlds—the physical realm, where one's feet touch the earth, one's ears capture sound waves, and one's eyes sieve light, and the realm of language, where one ceaselessly arranges symbols in particular patterns so as to connect with other beings who also move the same symbols in the same patterns.[1]

For all the complexity of each world, one has a certain primacy: the physical plane is the indispensable platform of the symbolic world. You articulate the abstract with the same throat, tongue, and lips that you use to eat, breathe, and taste. The buzz of your vocal cords resonates through all the bones of your body.

Because the two worlds overlap so much, people get to interact with other beings on both planes—with their bodies in the physical world, and within the language matrix. As I write these words—here at the Tea Lounge on Seventh Avenue in Brooklyn—you, the reader, may be anywhere—Manhattan, London, Melbourne—and I could be dead. It doesn't matter. In language, you and I are connected.

When you access the vast global network of computers we call the Internet, you can travel the world, find information, and interact with people in a way that was never before possible. The creation of the net was an awesome leap in technological evolution. Yet for all that it offers, it is the merest shadow of something much larger and much older. Language is the real information highway, the first virtual world. Language is the worldwide web, and everyone is logged on.

Introduction

In the popular imagination, evolution is a clean arc from rock hammer through arrowhead to Pentium processor. It's an inevitable outcome of one of those vague principles of life: the superior somehow unfolds out of the inferior. Likewise, language evolution begins with the urgent grunting of a guy with a club, moves through a "Me Tarzan, you Jane" phase, and ends finally with the epitome of civilization—the whip-crack enunciation of Sir Laurence Olivier as Hamlet.

For a long time the scientific account of language evolution wasn't that dissimilar to the popular version. Researchers sketched only a broad-stroke picture in which the complex somehow inevitably arose from the simple. Some thought, for example, that prior to *Homo sapiens'* current brilliance with words there existed a protolanguage, a clever form of communication that distinguished us in crucial ways from our fellow primates. But how did this protolanguage and the systems that followed it arise? Did a single genetic mutation shape the destiny of man? Or was it a slow layering of change over countless generations that propelled us from grunt to nominative case and from screech to sonnet—not to mention haiku, the *OED,* six thousand distinct languages, and words like "love," "fuck," "nothingness," "Clydesdale," and "aquanaut"? There are no easy answers.

Of all the formidable obstacles to solving this mystery, the first lies in the nature of the spoken word. For all its power to wound and seduce, speech is our most ephemeral creation; it is little more than air. It exits the body as a series of puffs and dissipates quickly into the atmosphere. On the evolutionary timescale, bone can last long enough to leave an impression, enabling us to track, for example, the adaptations that shaped 150 million years of ichthyosaurs. We can now see from the fossil evidence how these ocean dwellers changed over time, ballooning from a half meter into four-meter monsters, lengthening their spectacular snouts, and evolving fins and flukes from lizard bodies, before vanishing from the earth forever. But there are no verbs preserved in amber, no ossified nouns, and no prehistoric shrieks forever spread-eagled in the lava that took them by surprise.

Writing is a kind of fossil and so can tell us a little about the languages that have been recorded since it was invented. While it shares a lot with spoken language, including most of its words and much organizational structure, writing cannot be considered the bare bones of speech, for it is something else entirely. Writing is static, structured by the conventions of punctuation and the use of space. The kinds of sentences that occur in writing bear only an indirect relationship to the more free-flowing and complex structures of speech. Writing has no additional channels for avoiding ambiguity, as speech has with intonation and gesture. And writing is only six thousand years old.

In the absence of petrified words, evidence of change in language-related body parts offers a compelling clue to the course of language evolution. The brain, the tongue, the larynx, the lungs, the nose, and the uvula—the pendulous flap that swings in the throat of screaming Looney Tunes characters—are all intimately involved in speech production. But on the geologic timescale, soft tissue doesn't last much longer than a sound wave. It leaves traces only in very peculiar cases, like the skin of a thirty-thousand-year-old mammoth stalled in Siberian permafrost or the famous prehistoric iceman, a five-thousand-year-old mummy naturally preserved in an alpine glacier on the Italian-Austrian border.

For a long time the closest we could get to language-related fossils were the impressions left by the bones of distant ancestors. Scientists gained some useful information by interpreting cranial remnants, since skull size is an interesting, if indirect, measure of brain volume. Assumptions about the language skills of our forebears can also be made when considering the length of the neck vertebrae and the progression of other skeletal changes over time.

But the size of a skull or a femur takes you only so far. It doesn't tell you when the first word was uttered. Nor does it tell you if it was a noun like "tiger," a verb like "eat," or an imperative—"Run!" Bones can't tell you who said the first word or who was listening. Did language begin as a soliloquy, or is the fundamental nature of language to be communicative? A conversation requires at least two people—but how could someone invent language at exactly the same time that someone else figured out how to decode it? Fossils cannot answer this question.

Only very recently have scientists begun to work out how language evolved. But in the same way that no single fossil can provide an answer, no one researcher can solve this problem, which is fundamentally awesome and multifaceted. There will be no Einstein of linguistic evolution, no single grand theory of the emergence of language. Unearthing the earliest origins of words and sentences requires the combined knowledge of half a dozen different disciplines, hundreds of intelligent, dedicated researchers, and a handful of visionary individuals. Finding out how language started requires technology that was invented last week and experiments that were conducted yesterday. It also needs simple basic experiments that have never been done before.

It was only four centuries ago that electricity was discovered. A century later the principles of internal combustion were worked out and the engine was created. In the twentieth century we invented the computer, discovered DNA, split the atom, sent *Voyager 1* into the outermost regions of the solar system, and unraveled the human genome. It is only now that we are getting to the really difficult questions. Piecing together several million years of linguistic evolution without a single language fossil is not just a cross-discipline, multidimensional treasure hunt; it's the hardest problem in science today.[1]

This book tells two intertwined stories about language evolution. The first is an account of how it happened, the twenty-first century's best guess at humanity's oldest mystery: how the fundamental processes of evolutionary change spiraled together to produce an ape with protolanguage and, eventually, a linguistic primate—us. The second story is about what prompted a group of scientists to start asking questions about language evolution at this point in time. As with all scientific tales, it is a parable about humility and hubris, about progress and intellectual folly. It begins with a basic uncertainty about the validity of even studying how language evolved—think of this tale as the evolution of a doubt. The doubt began to nag at certain individuals thousands of years ago, and it has flourished over the years to concern many people in modern times. It has taken different forms: like Did language evolve at all? Is this question scientific? and Even if it is scientific, can we ever answer it? If you picked just one of these questions and followed its threads, you would soon find yourself tangled in a bloody thicket: Is the study of language a science? What counts as scientific evidence? Is language what makes us unique? What is language, anyway? What, in God's name, is science?

For me, this part of the story began in an introductory linguistics lecture in the early 1990s at the University of Melbourne. I can clearly remember my frustration when, after asking the lecturer about the origin of language, I was told that linguists don't explore this topic: we don't ask the question, because there is no definitive way to answer it.

That response made no sense to me—surely the origin of language was the central mystery of linguistics. After all, unlike any other trait, language is the foundation of our identity as individuals and as a species. As I later learned, the search for the origins of language was formally banned from the ivory tower in the nineteenth century and was considered disreputable for more than a century. The explanation given to me in a lecture hall in late-twentieth-century Australia had been handed down from teacher to student for the most part unchallenged since 1866, when the Société de Linguistique of Paris declared a moratorium on the topic. These learned gentlemen decreed that seeking the origins of language was a futile endeavor because it was impossible to prove how it came about. Publication on the subject was banned.

Today, nearly twenty years since my Linguistics 101 class, the field of language evolution is burgeoning. Important conferences on the topic occur regularly, with more announced each year. Journals are starting up, and books and collections of essays about language origins are published in increasing numbers. While fewer than one hundred academic studies of language evolution were published in the 1980s, more than one thousand have been published since.[2] Indeed, entirely new fields of science, like the digital modeling of language evolution, have been created.

Clues to the origin and development of the language suite have been found in areas as diverse as brain damage, the way that children speak, the way that chimpanzees point, and the genes of mice. Advances in the biology of language, artificial intelligence, genetics, animal cognition, and anthropology in the late twentieth century have shown scientists how previously uncharted mental and neural territory can now be explored. A lot of the research in these areas traces its roots to well before 1990, but since then there has been a winnowing around the question of language. What's more, an abundance of new evidence has been uncovered, introducing significant challenges to old methods and theories. In turn, this upwelling of interest and work has led to a greater synthesis between different projects in fields like linguistics, anthropology, genetics, and even physics. The fundamental goal of this book is to highlight with each chapter one of the moments or ideas that has given scientists compelling reasons to explore language evolution.

Part 1 of *The First Word* traces a broad historical arc through the lives of people intrigued by language origins, from King Psammetichus in ancient Egypt to Charles Darwin in nineteenth-century England. Darwin, of course, is best known for formulating the theory of evolution by natural selection, noting to great consequence that "the crust of the earth is a vast museum."[3] He had some very definite ideas about the evolution of language.

More recently, four figures bear much of the responsibility for the current state of the art. The first and most influential of these is Noam Chomsky, who went from being an exceptionally smart graduate student writing about the grammar of Hebrew to one of the most powerful intellectuals in history. The story of language evolution studies is unavoidably the story of the intellectual reign of Noam Chomsky. It is as much about his influence and what people think he said as it is about what he actually did say. Second is Sue Savage-Rumbaugh, who has taught a nonhuman how to use language. Third is Steven Pinker, the famous Harvard cognitive scientist who has written a number of influential and bestselling books about language and the mind. Finally, Philip Lieberman of Brown University, another cognitive scientist, started out at MIT as Chomsky's student and has since taken his experiments on language to the flanks of Everest and back. For this introductory history, think of Chomsky and Savage-Rumbaugh as poles in the debate, with Lieberman and Pinker somewhere in the vast middle. Between the extremes, there has been a collision of two completely different ways of seeing ourselves.

The first view, solidly anchored in popular linguistic theory, holds that language is a uniquely human phenomenon, distinct from the adaptations of all other organisms on the planet. Species as diverse as eagles and mosquitoes fly, whales and minnows swim, but we are the only species that communicates like we do. Not only does language differentiate us from all other animal life; it also exists separate from other cognitive abilities like memory, perception, and even the act of speech itself. Researchers in this tradition have searched for a "language organ," a part of the brain devoted solely to linguistic skills. They have sought the roots of language in the fine grain of the human genome, maintaining, in some cases, that certain genes may exist for the sole purpose of encoding grammar. One evolutionary scenario in this view maintains that modern language exploded onto the planet with a big genetic bang, the result of a fortuitous mutation that blessed the Cro-Magnon with the gift of tongues.

In the alternate view, a David to the Chomskyan Goliath, language is not a singular phenomenon or a specific thing. Rather, it is multidimensional—interdependent and interconnected with other human abilities and other cognitive tasks. Speech, for example, is crucial to language. And because we have a common ancestor, there is a strong family connection between our complex linguistic skills and the simple word and syntax skills that chimpanzees can acquire. Indeed, though our language system is unique, the progressive nature of language evolution also reveals an intimate relationship between our linguistic skills and the abilities of less closely related animals, like monkeys and parrots. Language is accordingly a higher cognitive function—one that emerges from multiple sites and operations in the brain. In this view, language is not a monolithic thing that we *have;* rather, it is a thing that we *do*. It arises from the coordination of many genetic settings; these are expressed as a set of physical, perceptual, and conceptual biases that underlie certain abilities and behaviors, all of which allow us to learn language.

Humans, in this view, are not so much unique and blessed geniuses, apes with that extra something special. Rather, we are special but also in key respects very much the same—which makes us not so much a higher species as the earth's idiots savants, narrowly and accidentally brilliant, juggling symbols like there is no tomorrow and nothing else to do. Some of the most radical proponents of this perspective think of language itself as an organism, one that evolves to suit its own needs.

The first perspective has dominated linguistics and cognitive science since the middle of the twentieth century. Since the 1990s, as researchers in different disciplines all over the world have grappled in earnest with language evolution, many have found that they are converging on the second. But instead of flipping entirely from one view to the other, the field is building a less absolute but more satisfying body of knowledge. The David-Goliath clash has not so much been lost or won as transformed into a struggle that is much more complex, though still deeply felt. What is both marvelous and perplexing about this struggle is that the resolution of one mystery often gives rise to another. For example, it now seems true to say that language arose very recently. It is also equally true to say that it did not. It's reasonable to claim that human language is unique. But it is also useful, interesting, and fair to say that human language lies on a continuum that includes other human abilities and the abilities of nonhuman animals. As you'll see throughout this book, language itself is one of the biggest obstacles to clarity in the study of language evolution.

After this tour of important historical moments, part 2 asks how language evolved. Or rather, it doesn't, explaining instead that that question is simply too monumental. What's more, it is misleading. The word "language" is used to describe too many different phenomena, like the words we speak, the particular language we speak, the universal features all languages have in common, and the suite of inclinations and capacities that enables us to learn the language of our parents. Part 2 outlines and examines the language suite—what abilities you have if you have human language. It takes the sounds we make, the way we string them together, our interactions and the learning in speech, and our gestures, imitation, and genes and explores when each of them appeared in the evolutionary trajectory of human beings.

Even though the comparative, genetic, and linguistic evidence of part 2 demonstrates how various aspects of the language suite evolved so that we were born able to learn the language of our parents, all this amazing research leaves one crucial question unanswered: How did the language of our parents get here in the first place? Part 3 examines this question. It asks how all the pieces of the language suite wound together over time to give us what we have today. It introduces young researchers like Simon Kirby in Edinburgh and Morten Christiansen at Cornell, who use computer models to show that language can evolve all by itself. Kirby and Christiansen argue that one of the most useful ways to think of language is as a virus, one that grows and evolves symbiotically with human beings—meaning that language shifts around and adapts itself in order to develop and survive.

Part 4 looks at what happens next. I wrote earlier that the study of language evolution had boomed since I was a linguistics undergraduate, and in fact the tumult and commotion have increased even more since I started writing this book. The debate about how and why language might have evolved was rejoined ever more loudly in 2002, when Noam Chomsky first published on the topic. Part 4 also ponders the future of language. If language has evolved, where will it head next? And where, for that matter, will we? New scholarship that claims the human species has stopped evolving biologically is discussed.

Part 4 also examines why evolutionary narratives have been so unpopular in the field of language evolution. Finally, the epilogue asks what would happen to language if you shipwrecked a boatload of pre-linguistic babies on the rocky shores of Galápagos.

This book is shaped by the fact that reporting on the life of an idea is a slippery task. If it were simply a matter of trying to render the intangible tangible, it would be hard. But it's made more difficult by the fact that the abstract doesn't exist, so to speak, in the abstract. Ideas are frustratingly anchored in the heads of individuals, and each of those individuals has his own version of any one thought. They all agree on some of the implications and none of the others. And everyone has a slightly different set of assumptions, not all of which he is conscious of or willing to admit to.

Additionally, even though ideas come from nowhere but the heads of people, attributing them to individuals is tricky. Most ideas have been around forever in some form or other, yet the tides of thought follow no clear pattern. An idea can lie neglected for a millennium and then suddenly become invigorated by the agenda of a new age. And even when someone does come up with a really novel thought, it's inevitably the case that someone else across the Atlantic or the Pacific has awoken that very morning with the same bulb flaring above her.

We all want to believe that ideas rise or fall on their own merits, but in the real world they don't. Both personality and ideology shape the pursuit of knowledge and affect the way an idea gets lost and found over the years. And if this is not just, then perhaps it is natural. After all, what we're finding is that culture—which at its most basic is an interaction between two individuals—is a great force of evolution. Our personalities, our ideologies, and our ideas all arise from the same place—the intersection between biology and experience. Is it any wonder they are inextricable?

The intent in telling this story is to render the large shifts of history. As much as possible, I try to avoid the hindsight that makes thinkers seem more sophisticated than they probably were at the time. I stay as far away as possible from labels like "behaviorism" and "positivism." If you use these words precisely, each requires a manifesto of explication and qualification. Alternatively, if you use them but forgo the provisos, they tend to make people and ideas seem like caricatures. (For an intensely subtle history that tweezes apart each word in many historical utterances read *Grammatical Theory in the United States from Bloomfield to Chomsky* by Peter Matthews[4] or *The Linguistics Wars* by Randy Allen Harris.)

The study of language evolution has been enacted by a cast of hundreds. If you try to isolate the really gifted ones and the ones who represent a particular idea and the ones who'll give you great quotations, there are still too many people to describe. This book focuses on the individuals that it does because they have made an idea their own in a striking and significant way. Perhaps, like Steven Pinker, they have combined a genius for cognitive science with a genius for timing. Or like Simon Kirby, they occupy a historically unique spot. Kirby, a young professor of language evolution at the University of Edinburgh, is not only spearheading research into the digital modeling of language; he was the first student to ever take an undergraduate and a graduate degree majoring in the evolution of language.

Some events in this book are retold by just one individual; some are recounted by more. Other happenings and thoughts have been bequeathed to a kind of large, collective memory. As much as possible, I have tried to be clear about who is doing the remembering and whose lens is providing the viewpoint.

Noam Chomsky stands out in this book as a hugely influential figure. He is also an abominably difficult subject. His theories and terminology have changed many times since the 1960s, and there are no complete and reliable road maps to these shifts. This drives academics as crazy as it does writers.

Every interview I conducted for this book left me excited and always leaning toward the particular theory of whomever I had just spoken to. Everyone I interviewed was dynamic and engaged; a few were modest as well as intelligent. While all of them believe in science as the pursuit of truth, they also treat science as a competition. This eclectic group of psychologists, biologists, neuroscientists, and linguists (and the hundreds of people who are here but unnamed) are reaching into the deep past to crack open a mystery more than six million years in the making. Keep in mind that some of them will turn out to be completely wrong.

Naturally, everyone thinks he has the right solution. At the time I wrote this introduction, pretty much every one of the main characters in this book, and a slew of others, was writing his own book to present at greater length his particular version of how language evolved.

Why does language evolution matter? Because the story of language evolution underlies every other story that has ever existed and every story that ever will. Without this one tale, there are simply no such things as beginnings, middles, and ends. Only because the evolutionary plot unfolded in the way it did do we have yarns, fables, and parables, tragedies, farces, and thrillers, news reports, urban legends, and embarrassing anecdotes from childhood. The ultimate goal of this book is to present fragments from an epic about an animal that evolved, started talking, started talking about the fact that it was talking, and then paused briefly before asking itself how it started talking in the first place.

I. LANGUAGE IS NOT A THING

Prologue

The Panthéon in Paris sits on a hill, and when you stand on its roof and look out from each corner, the City of Light stretches beneath you. To the northwest the Eiffel Tower stands on the skyline, enormous but light. The Panthéon, in contrast, is massive. Its main chamber is supported by huge stone columns, and one floor below it lies the crypt. It's damp and dark, and among its many graves is the one belonging to Jean-Jacques Rousseau, who died in 1778.

Although he is best known for his political philosophy, Rousseau spent considerable time thinking about where language came from and what humans were like before they acquired it. He imagined a primeval forest where people wandered alone. If males and females crossed paths, they'd pause for sex, then go on their separate ways. Mothers and children abandoned one another as soon as possible, and because proximity was the only way they could tell they were related, a brief period apart soon rendered them unable to recognize their own kin.

In his *Essay on the Origin of Languages,* Rousseau wrote that when these roving, isolated creatures did communicate, they used crude cries and gestures that imitated animal vocalizations. A barking sound, for example, meant "dog." Eventually, the bark came to represent the animal, and as humans expanded their repertoire of animal mimicry, they created the first words.

This primal lexicon was limited by the range of sounds that could be mimicked, so Rousseau suggested that the original language was mostly gestural. Hand and body movements didn't just supplement meaning; they formed an imprecise kind of sign language that worked in tandem with the vocal pantomime. Rousseau believed that language originally burst forth in times of crisis: "Man's first language, the most universal, the most ener-getic and the only language he needed before it was necessary to persuade assembled men, is the cry of Nature . . . wrested from him only by a sort of instinct on urgent occasions, to implore for help in great dangers or re-lief in violent ills."[1]

We eventually lost this body language, theorized Rousseau, because gestures aren't as versatile as the spoken word. Hand movements can't be seen at night or if the line of sight is somehow blocked.

The eighteenth century was a time of energetic conjecture about the evolution of language, and Rousseau was heavily influenced by thinkers like the German Romantic philosopher Johann Gottfried von Herder and Étienne Bonnot, the abbé de Condillac, who wrote a number of treatises on the nature and origins of language. Herder believed that humans first mim-icked animal sounds to communicate. He also thought they imitated other sounds of nature, like the rustling of the wind or the babble of water.

All of these ideas had a lasting impact on the way we think about lan-guage evolution, and in some form they're still taught in many classrooms today. They are explained, and caricatured, as the "bow-wow" theory of evolution. The theory goes like this: If you told a stranger on the street, "There's a mad dog a block away," these few words should be enough to save him from a possibly dangerous encounter. If you discovered the stranger spoke another language, your warning might take longer to con-vey, but you'd still try, and you'd probably be successful. Waving hands and rolling eyes might do it, and a natural flair for the dramatic would help. But to really sell the message, you should bark like a dog while jabbing your finger in the correct direction.

The ease with which humans produce and understand such signals when words don't work—the combination of gesture and auditory mim-icry—feels innate. Proponents of such a theory would argue that not only is it innate but it's how we communicated before we had language and it was, in fact, crucial to the evolution of language.

Other general ideas about how language evolved that have been bandied about for ages—and, like the bow-wow theory, actually explain very little—include the "yo-he-ho" and "pooh-pooh" theories. "Yo-he-ho" stands for the rhythmic grunts and chanting of people working together, and according to the theory it is from such social cooperation that language arose. The "pooh-pooh" theory proposes that language originated in cries of emotion. (Perhaps it would more accurately be called the "ouch" theory of language evolution.)

Rousseau is a key representative of an important period in language evolution, standing at the brink of modern thought and theorizing. But his era was not the first in which men began to question where words came from. Several millennia before Rousseau was born, the Pharaoh Psammetichus believed that one single language must have been the source of all subsequent human languages. In search of that first tongue, the ancient Egyptian king isolated two babies in a mountain hut. He sent a shepherd to feed and clothe the children but forbade the man to speak to them.

It was thought that with no exposure to speech, the original human language would emerge from the children's mouths as naturally as their hair would grow and their limbs would lengthen. According to the Greek historian Herodotus, the children eventually uttered the word *bekos,* which is Phrygian for bread, which led Psammetichus to deduce that Phrygian was the first human language, even though it was not his own.

The quest for humanity's mother tongue spans centuries and cultures. Seekers believed there once existed a "monolinguistic golden age." Rediscovering the first ancient tongue was considered a way to re-create this time, a chance to attain perfect expression by conveying one's thoughts and intentions without ambiguity.

The pharaoh's test was repeated at least twice: The Holy Roman Emperor Frederick II of Hohenstaufen, a soldier, diplomat, and scholar who was said to be fluent in Latin, Arabic, Hebrew, French, German, Italian, and Greek, likewise confined two children to silence in the early Middle Ages. His subjects died before articulating a word. Two hundred years later, King James IV of Scotland imprisoned two Scottish children who in the end, apparently, "spak very guid Ebrew."

The Church held for centuries that Hebrew was the first language, but scholars proposed many other contenders. In the fifteenth century, the architect and scholar John Webb argued for the supremacy of Chinese, claiming that the biblical Noah had washed up in China after the flood.

Chinese remained a popular candidate for a few centuries, with Joseph Edkins writing in 1887 in *The Evolution of the Chinese Language* that it had to be the world's primeval language simply because of its age. Noah Webster proposed in 1830 that the primordial language was Aramaic, another Semitic language and the native tongue of Jesus.

Inseparable from the notion of a single tongue that united all humanity is the idea that language is exclusively a property of human beings and one that originated with the source of all life. Before the Darwinian revolution, it was thought that there was no prehuman existence and no pre-linguistic human experience. Consequently the first acts or expressions of language were universally said to be divinely inspired. In other cultures and times, the Egyptians believed the god Thoth was the progenitor of language, while the Babylonians attributed it to Nabu. For the Hindus, Sarasvati, wife of Brahma (creator of the universe), gave language to humanity.

When Rousseau and his fellow thinkers imagined a world before words, they pictured an extended period of language genesis. Instead of being a magical property of humanity, language was something our species acquired over time. This new model of evolution shifted the focus from a perfect first language from which all varieties descended to language as undergoing a developmental stage that resembled the communication systems of other animals. Even though Rousseau is well known as a believer in the unbridgeable line between humans and the rest of creation, this shift left the sharp division between us and the rest of the animal world a little blurred.

Proposing the existence of more than one stage of language development immediately raises the issues of how and why people moved from one stage to the next. What forces drove us to speak in the first place? What passions shaped the way language was formed?

Although Darwin mentioned language very little in *On the Origin of Species*, the book is a keystone for every discussion about language evolution that has followed it. In fact, all debate about who we are and how we came to be on this planet can be divided into conversations that took place before publication of *Origin* and those that have taken place after it. *Origin* was printed six times during Darwin's lifetime, and many times since. Not only did it introduce the concept of evolution (truly the most superlative-laden theory in science; Jared Diamond's evaluation—"the most profound and powerful idea to have been conceived in the last two centuries"[2]—is typical), but it initiated the modern study of evolutionary biology. The flow of books published about Darwin every year seems endless.

Darwin focused more on language in *The Descent of Man* (1871) than in *Origin*. Language was not a conscious invention, he said, but "it has been slowly and unconsciously developed by many steps."[3] At the same time, he noted, humans don't speak unless they are taught to do so. Psammetichus's experiment could never have worked, because language is "not a true instinct."[4]

Darwin believed that language was half art, half instinct, and he made the case that using sound to express thoughts and be understood by others was not an activity unique to humans. He cited the examples of monkeys that uttered at least six different cries, of dogs that barked in four or five different tones, and of domesticated fowl that had "at least a dozen significant sounds." He noted that parrots can sound exactly like humans and described a South American parrot that was the only living creature that could utter the words of an extinct tribe.[5] Darwin included gesture and facial expressions under the rubric of language: "The movements of the features and gestures of monkeys are understood by us, and they partly understand ours."[6]

He also considered animals' abilities of comprehension and cognition. "As everyone knows," he wrote, "dogs understand many words and sentences." He likened them to small babies who comprehend a great deal of speech but can't utter it themselves. Darwin quoted his fellow scholar Leslie Stephen: "A dog frames a general concept of cats or sheep, and knows the corresponding words as well as a philosopher [does]."[7]

Darwin also pointed out compelling parallels between human language and birdsong. All birds, like all humans, utter spontaneous cries of emotion that are very similar. And both also learn how to arrange sound in particular ways from their parents. "The instinctive tendency to acquire an art," said Darwin, "is not peculiar to man."[8]

Where humans differ from other animals, Darwin believed, is simply in our greater capacity to put together sounds with ideas, which is a function of our higher mental powers. What got us to that level was love, jealousy, triumph—sex. Before we used language as we know it today, we sang, producing "true musical cadences" in courtship.

On the Origin of Species and, in more detail, The Descent of Man also discuss similarities in the way that languages, like animals, change over time. Just as species split off from one another to form new groups, languages split to form dialects and entirely new languages. From the common ancestor of all mammals, many different species arose, like the manatee, the horse, and the gorilla. Likewise, Latin branched over time to give rise to the modern Romance languages, including Italian, French, and Spanish.[9]

Darwin's theory of language change was embraced most enthusiastically by scholars of language. Evolutionary theory turned out to be a perfect analogy for language phenomena that they had observed but were unable to account for in any systematic way. Linguists of the nineteenth century (known as philologists) are often described as having been overly preoccupied with their status as genuine scientists, and their newfound ability to explain language change in terms of biology and natural history gave them a greatly desired sense of credibility.

Biological evolution proved to be an excellent analogy for language change, and linguists took up the evolutionary analogy with such enthusiasm that they began to treat natural selection as a literal account of language change rather than as a helpful analogy, applying the idea of survival of the fittest to such phenomena as the ways that speech sounds change over long periods of time (how, for example, a distinct sound like f might become s). Ironically, linguists still regarded speculating on the origins of language to be an unscientific problem, and it remained controversial to adopt Darwin's theory for that purpose. So while Darwin himself freely considered the origins of language, linguists did anything but.

The distaste for speculation about language origins culminated in an extraordinary move by the Société de Linguistique of Paris in the nineteenth century: it banned any discussion of the subject, even though it was attracting more and more attention. Its pronouncement read: "The Society will accept no communication concerning either the origin of language or the creation of a universal language."[10] In 1872 the London Philological Society followed suit.

This act of academic censorship had remarkably long-reaching consequences. Despite the occasional flare of interest, language evolution was considered a disreputable pursuit for more than a century. In 1970 a meeting of the American Anthropological Association presented a number of papers on language evolution, many of which were later collected in the book *Language Origins*. Even then, a contributing anthropologist wrote that scholars who studied the subject did so either apologetically or with reluctance.[11] In 1976 the New York Academy of Sciences collected another series of conference papers on the topic in a volume called *Origins and Evolution of Language and Speech,* and in 1988 proceedings from a NATO summer institute organized by Philip Lieberman were published. The volume was called *Language Origin: A Multidisciplinary Approach.*[12] Yet despite the widespread interest that these collections suggest, the field remained marginal. This changed in the 1990s with the publication of one article about language evolution that drew together commentary from researchers with dramatically different ideas of what language is. Since then, tensions between the types of research, and researcher, have energized the topic, causing it to finally flourish.

1. Noam Chomsky

Housed in the modern, gabled, jarringly chrome, brick, and mustard yellow Stata Center at MIT is the Department of Linguistics and Philosophy. Noam Chomsky has had an office in the department for forty-five years. His room is full of shelves with books, five rubbery office plants, and a small table in the center facing a poster of Bertrand Russell. Under Russell's looming face is the quotation: "Three passions, simple but overwhelmingly strong, have governed my life: the longing for love, the search for knowledge, and unbearable pity for the suffering of mankind." Across Chomsky's desk stretch piles and piles of books and unbound manuscripts. They look like a small mountain range.

Prior to an office interview, Chomsky spoke at the 2005 Morris Symposium on the Evolution of Language at Stony Brook, New York. There, his speech seemed flat, almost without affect. He stood at a lectern and read directly from a paper, speaking in such low tones that it was sometimes hard to make out what he was saying. Today, in person, he accompanies his greeting with a puckish grin but is otherwise grave. He takes a seat at the table and sits very still, talking in such a forceful stream that it is virtually impossible to get a word in edgewise. The sense that he cares deeply about what he is saying is unmistakable and compelling.

Chomsky's style of exposition in person is almost exactly the same as in his writings—he takes no prisoners. Depending on whether you disagree or agree with him, you will probably experience his manner as one of airless conviction or the just impatience of a man who knows the truth and is weary of waiting for others to get it. Debating him is a high-stakes venture—he shows little respect for the intelligence of those who don't accept his views.

Chomsky has served as a geographical constant in the minds of generations of scientists and linguists since the early 1960s. It was as if, on the publication of his first book, he thumped down a flag and said, "This is the North Pole," and the rest of the scientific world mapped itself accordingly.

Anyone who has studied language or the mind since then has had to engage at some level with Chomsky's definition of language. Chomsky's signature claim is that all humans share a "universal grammar," otherwise known as UG, a set of rules that can generate the syntax of every human language. This means that apart from the difference in a few mental settings, English and Mohawk, for example, are essentially the same language. Traditionally researchers committed to Chomskyan linguistics believed that universal grammar exists in some part of our brain in a language organ that all humans possess but no other animals have. For Chomsky, syntactic structure is the core of human language, and a decades-long quest for the universal grammar—the linguistic holy grail—has shaped linguistics since he first presented his ideas.

Around the time of the Stony Brook conference, the British magazine *Prospect* published the results of a poll in which Chomsky was voted the world's top intellectual. (He beat Umberto Eco, who took second place, and Richard Dawkins, in third.) Twenty thousand voters, mainly from Britain and the United States, had been canvassed, and a flurry of media about Chomsky had accompanied the poll's announcement. *Prospect* published two articles about the world's top intellectual: a "for" and an "against" Chomsky. On the "for" side Robin Blackburn wrote that Chomsky had transformed an entire field of inquiry and likened him to the child who pointed out that the emperor had no clothes. On the "against" side Oliver Kamm spoke of Chomsky's "dubious arguments leavened with extravagant rhetoric."[1]

This latest burst of attention is merely one of many. Chomsky has been famous in several worlds for a long time. Within the university there are apocryphal Chomsky stories. It's said that graduate students would sometimes come to their meetings with him in pairs, so they could take turns, trying to keep up. His weekly seminars are legendary. Over the decades, they have been attended not just by MIT graduates but also by an ever-changing cast of unfamiliar students, whom none of the regulars knew. Time and again, so the story goes, the outsiders would try to beard the lion in his den, and Chomsky would swat them one by one. By now, it has to have become tiresome.

Until 2002, and in some ways even since then, Chomsky's exact position on the evolution of language was hotly contested, but both sides in the debate would at least agree on this: for many years Chomsky deemed language evolution unworthy of investigation, and given the extraordinary nature of his influence, his pronouncement was as deadening as any formal ban. Now, he has decided, it is feasible to study the topic.

Before Chomsky, most linguists were field linguists, researchers who journeyed into uncharted territory and broke bread with the inhabitants. They had no dictionary or phrase book but learned the local language, working out how verbs connect with objects and subjects, and how all types of meaning are conveyed. They have always been seen as adventurers, but the soul of a field linguist is really that of a botanist. When they transcribe a language for the first time, they create a rigorous catalog of sounds, words, and parts of speech, called the grammar of the language. Once this is completed, they match one catalog to another—finding evidence of family relationships between languages. Grammar writers are meticulous and diligent, arranging and rearranging the specimens of language into a lucid system.[2]

In the early 1950s, Chomsky submitted a grammar of Hebrew for his master's thesis at MIT. At the same time he was also at work on a huge manuscript titled *The Logical Structure of Linguistic Theory,* in which he wrote about grammar in the abstract.[3] Instead of describing an actual language, Chomsky discussed the different ways that a language *can* be described. He submitted one chapter of this effort for his Ph.D. thesis, but it was so different from the way linguists typically thought and worked that many academics who read it didn't really know what to do with it.[4] In 1954 Morris Halle, an MIT professor famous for his work on the sounds of language, wrote to Roman Jakobson, another famous linguist: "I am very impressed with Noam's ability as a linguist; he has a wonderful head on his shoulders, if only he did not want to do all things in the most difficult way possible."[5]

With his next project Chomsky moved even further away from the concerns of his colleagues. After receiving his doctorate, he got a part-time job at the Research Laboratory of Electronics at MIT.[6] He carried on with his work, taught linguistics, and, in order to make enough money, also taught German, French, philosophy, and logic. In 1957 Chomsky published the notes from his first linguistics course as *Syntactic Structures*.

In that book he continued his examination of language in the abstract, discussing the grammars of languages in a wholly new way. Instead of simply being a catalog of all the words and sounds in a language, with instructions for how to put them together, a grammar, he argued, was really a theory of that language.

As a theory, a grammar should be judged in the same way all scientific theories are: it should explain as much as it can with as little as possible. It should be simple and elegant. Viewed this way, possible grammars of a language can be compared in the same way that different theories in science are: the successful one more fully explains the phenomena in question in as economical terms as possible.

Syntactic Structures, for example, contrasted two methods for writing a grammar. The best method, said Chomsky, collapsed all of language into a set of rules. And in much the same way that software generates output in a computer, those rules can generate an entire language. For example, an English sentence can be described as "S goes to NP VP," meaning that a sentence (S) consists of a noun phrase (NP) and a verb phrase (VP). "NP goes to Det N" means that a noun phrase consists of an "a," the determiner (Det), and a noun (N).[7]

Chomsky also pointed out that the set of language rules could be made smaller and simpler if you included ways to relate certain sentences to each other. "The man read the book" and "The book was read by the man," for example, have a striking similarity. Instead of having separate rules for each of them, Chomsky suggested that the more complicated second sentence was derived from the first. He called this a transformation.[8]

If the phrase structure analysis of "The man read the book" is "S goes to NP_1 VP NP_2," then "The book was read by the man" can be represented as "S goes to NP_2 VP by NP_1." In this way, the relationship between all the simple active sentences of English and their passive versions can be described by just these two simple structures and the transformational rule that links them.

Language, in this view, is basically a set of sentences. And the job of a grammar, or theory of language, is to generate all of the language's allowable sentences ("The cat sat on the mat"; "The plane was rocked by turbulence") but none of the bad ones ("Cat mat the on sat"; "Turbulence plane by the rocked was"). A grammar generates all possible utterances of a language, Chomsky said, "in the same way that chemical theory generates all possible compounds."[9]

Syntactic Structures got Chomsky some attention, but at the time of publication it wasn't especially well known. Two years later Chomsky made a much larger splash when he published a review of B. F. Skinner's *Verbal Behavior*. The review appeared in what was at the time the premier journal of linguistics, *Language*. Skinner, a psychologist, was already well known for his theory of behaviorism. In its simplest form, behaviorism says that all animals, humans included, are like machines—if you press their buttons in the right way, they'll respond automatically. The appearance of emotion or thought is irrelevant, because everything can be reduced to behavior. As long as you know what kind of machine you are dealing with—human, feline, avian—you can control its behavior. Even very complicated behavior can be reduced to a series of depressed buttons.

At the time, people spoke about Skinner in the terms they would later use to describe Chomsky. In her book *Animals in Translation,* Temple Grandin wrote about the behaviorist's influence when she was a college student. "Dr. Skinner was so famous," she remembered, "just about every college kid in the country had a copy of *Beyond Freedom and Dignity* on his bookshelf." Of behaviorism she added, "It's probably hard for people to imagine [the power] this idea had back then. It was almost a religion. To me—to lots of people—B. F. Skinner was a god. He was a god of psychology."[10]

Chomsky's review was published two years after Skinner's book came out, oddly late in the day for a book review, even in academia. Nevertheless, it had an immediate impact. Skinner suggested that language was a simple behavior, a notion Chomsky dismissed as absurd. Skinner was used to dealing with lab rats, but pressing a pellet for food is no analogy for producing language. In order to speak, people use great creativity while obeying many complicated rules.

Chomsky argued: "A typical example of stimulus control for Skinner would be the response to a piece of music with the utterance *Mozart* or to a painting with the response *Dutch*. These responses are asserted to be 'under the control of extremely subtle properties' of the physical object or event." But, argued Chomsky, what if we don't say "*Dutch*"? What if we say, "*Clashes with the wallpaper, I thought you liked abstract work, Never saw it before, Tilted, Hanging too low, Beautiful, Hideous, Remember our camping trip last summer?* or whatever else might come into our minds when looking at a picture"? People are not controlled by some unknown aspect of a painting, he said. Their response comes from inside them and is facilitated by the infinite creativity of language.[11]

The key idea in Skinner's behaviorism—if you push someone or something in the right way, it will respond in a predictable manner—was called stimulus-response. But when it comes to language, Chomsky said, particularly when children learn language for the first time, stimulus-response is not a relevant model. What is fundamentally interesting about language is the incredible speed with which children learn thousands and thousands of words and the many rules that combine them. In fact, there just isn't enough information in the language children hear in their day-to-day lives for them to divine all the rules that they come to know how to use. Chomsky called this phenomenon "poverty of stimulus." So how do children learn how to speak if language is so incredibly complicated? They must come to the task somehow prepared, he concluded. They must be born with a mental component that helps them learn language.

It was as if Chomsky had delivered unto Skinner and behaviorism a knockout punch.[12] The review garnered enormous amounts of attention from people in all sorts of disciplines. For many academics, this was the moment at which Chomsky seized their attention and would hold them riveted from then on.

The young professor was propelled into the limelight, and even though his review was widely criticized as glib, biting, and angry, it was these very qualities that seemed to thrill people. As much a polemic as a review, the article was described as "devastating," "electric," and a superb job of "constructive destruction." Chomsky the linguistic freedom fighter was born.[13]

Skinner responded that Chomsky hadn't understood what he was saying, that in some respects it seemed that Chomsky had intentionally misinterpreted him, but the damage was done. From that point on, the obvious influence of behaviorism seemed to fade.

It took a few years for the impact of Chomsky's first book to be felt, but by 1964 Charles Hockett, one of the most eminent linguists of the time, described *Syntactic Structures* as among the field's few "major breakthroughs."[14] Howard Maclay wrote: "The extraordinary and traumatic impact of the publication of *Syntactic Structures* by Noam Chomsky in 1957 can hardly be appreciated by one who did not live through this upheaval."[15] Ray Jackendoff remembers that in 1965, when he began his graduate studies (with Chomsky), "generative linguistics was the toast of the intellectual world."[16] Daniel Dennett, the well-known philosophy professor at Tufts, wrote in *Darwin's Dangerous Idea* that he could "vividly remember the shockwave that rolled through philosophy when Chomsky's work first came to our attention."[17] Looking back, Chris Knight of the University of East London wrote that Chomsky may as well have thrown a bomb.[18]

In less than a decade, people were proclaiming a psycholinguistic revolution.[19] Many young scholars flocked to MIT to work with Chomsky on his new generative linguistics, and in many other universities researchers began to search for the mental component containing the basic, innate generative rules of language with which children are born.

Chomsky's theory was expanded and his reputation solidified with *Aspects of the Theory of Syntax,* published in 1965. A slim but extremely difficult book, *Aspects* further explained key Chomskyan concepts like deep structure and surface structure, and has since become a classic text.

All the ideas in *Aspects* rest on the notion that language can be divided into, on the one hand, everything that goes along with actually speaking in a given situation and, on the other, all that is stable and universal. Chomsky called this the difference between competence and performance. Competence, which includes syntax (a perfect, mathematical system), is the innate basis of language and is the same from speaker to speaker. Performance includes whatever is individual or context-specific in language: the myriad differences in the way we pronounce "ketchup," the use of gesture, the "ums" and the "ahs."

Even though he imagined an idealized speaker and hearer as the subjects of his research, language in the Chomskyan sense had little to do with the fact that it overwhelmingly takes place between people. For the Chomskyan linguist, to study what was interesting about language was to discard any variation, the way any given speaker actually speaks, and to focus instead on the skeleton that remains.

The role of the language specialist was fundamentally changed by these ideas. Linguists were no longer mere catalogers but scholars who were perfectly positioned to unearth the deepest mysteries of their subject. What mattered about a language was not that it came from a particular region like the plains of the Midwest, the villages of Mexico, or the beaches of Asia but that it came from our heads. With generative linguistics, the terrain that the linguist explored shifted from the corners of the planet to the depths of the human mind. Universal grammar specified every rule for every language, and that controlled a child's ability to develop the correct rules of syntax of each language. It was believed in the early days that universal grammar, or the language organ, was hardwired into people's brains. Anyone born with UG, which is to say everyone, was born with the potential to learn any language.

Even though searching for the universal principles of language was hugely different from the way scholars had previously thought about language, early generative linguistics still divided language in the brain in much the same way that linguists of the 1950s had divided languages in the field. Field linguists wrote a grammar by analyzing its structure, sound, and meaning in separate sections. They also believed that when you were learning a language from scratch and assembling its grammar, you should keep these parts of language completely separate—you should never mix levels.

Generative linguists began to divide language in the brain in the same way. They looked for evidence of a module that controlled syntax, a module that controlled meaning, and a module that processed sound. It was thought that these modules were independent of one another and that language was produced by a coarse-grained interaction between them. Additionally, the separate systems of language had their own subsystems. For example, the syntactic module was made up of a set of smaller modules, each dealing with a different part of syntax, each autonomous.

In this model, when someone heard speech, the separate modules divided up the signal. The syntactic module extracted from the sound wave all the information regarding syntax, the intonation module analyzed all the pitch variation, and so on. Once each module had sufficiently analyzed the component for which it was responsible, the brain put them all back together as language. One implication of this theory is that when you heard someone else speak, the grammar part of your brain somehow extracted the grammatical information from the sound waves but ignored any other information in those waves that might help interpret it.

The workings of the language organ were also thought to be completely separate from other parts of the brain. They were separate from the context of spoken language, and they were also completely different from similar systems, like music. Gesture was peripheral and uninteresting. Moreover, human language was entirely distinct from the communication that takes place between other animals. This model of language was consistent with general theories at the time about how the brain functioned—namely, as a series of separate boxes, each of which computes different parts of the world.

Critics said the model was merely a new version of phrenology, a nineteenth-century "science" that held that for every tendency in an individual, there was a corresponding spot on the brain that controlled it. The brain would bulge or recede in these areas, depending on how developed a given trait was. (Phrenologists even believed that the skull would echo the shape of the brain, so that a person's character could be read by the bumps and pits of his or her head. For example, someone with a great deal of self-esteem would have a big bump right at the top and back of her head. Phrenology is now the iconic example of silly science.)

The Chomskyan deconstruction of language was, on the one hand, counterintuitive. The average person who hadn't taken a university course in linguistics and been rigorously trained to force these elements of language apart would probably consider context crucial to understanding language. He would count intonation as important, and he would be unlikely to completely separate structure from meaning.

Yet Chomsky's approach satisfied another kind of intuition: to divide an object into its essential and incidental parts. With language, generative linguists tried to strip away everything peripheral, anything that *could* be stripped. The hope was to expose the bare bones, discover what was indivisible, and unearth the core.

Another key insight that Chomsky brought to language studies was the infinitude of language. While so much of language is rote, consisting of things that you have heard before, you don't have to go far to find words assembled in a way you've never heard them put together. Chomsky described this as the infinite use of finite means, calling it "discrete infinity."

With discrete infinity, "Kate read the book that Bill wrote" can be embedded in "Ally saw," becoming "Ally saw that Kate read the book that Bill wrote." It can be further embedded into something like "Andrew explained how Ally saw that Kate read the book that Bill wrote," and so on, ad infinitum.

Ten years after *Syntactic Structures* was published and two years after *Aspects,* most papers presented at the 1967 meeting of the Linguistic Society of America discussed Chomsky's transformations.[20] A few years later Chomsky's growing reputation within linguistics and philosophy had spread into many other fields. In 1970 a Chomsky monograph was published in the Viking Press Modern Masters series, putting him in the company of Einstein and Freud.

Of course, Chomsky had detractors at this time as well, and the louder his supporters became, the more his critics grew in number. In 1967 Charles Hockett, who had just three years earlier hailed Chomsky's genius, called him a "neo-medieval philosopher." Another prominent linguist, George Trager, described him a year later as "the leader of [a] cult . . . with evil side-effects."[21]

Chomsky's skirmish with B. F. Skinner turned out to be merely the first in a long line of infamous, bitter conflicts. The next took place in the late 1960s and early 1970s, when a group of linguists calling themselves generative semanticists argued that separating language from the way it was used was ridiculous.[22] This group believed that the most fundamental organizing principle of language was its meaning (semantics), not the way it was structured (syntax), as Chomsky's transformational theorists believed.

The generative semanticists defined themselves in opposition to the Chomskyan juggernaut, and as Randy Allen Harris (the main historian of this period) recounted it, that opposition took on all the flavor of the 1960s counterculture—irreverent, exuberant, and combative. Their criticisms of Chomsky extended from the way that he divided up language to his ascetic style. One running joke of the era was inventing a title for the world's shortest book, like *"Problems of the Obese" by Twiggy*; a popular candidate among linguists was *The Bawdy Humor of Noam Chomsky*.[23] In turn, the generative semanticists were caricatured as unthinking followers of a fad. Chomsky repeatedly insisted that they didn't actually understand the theories with which they took issue.

There is a clear pattern in these different conflicts. Again and again, Chomsky's critics claimed that he chose data to support his theories but then discarded it when it no longer suited, and that he intentionally misinterpreted his adversaries and then launched an attack against his own misunderstanding. People also accused him of abandoning ideas that he once promoted without acknowledging that he had changed his position. Another complaint was about the way Chomsky dealt with counterevidence to his theories, most of which he insisted could be simply disregarded.

When Chomsky put forth his ideas, he typically dictated the terms with which people could reasonably disagree with him. Academics objected to the fact that he laid out his argument and the rules for argumentation at the same time. For instance, he said, "Counterexamples to a grammatical rule are of interest only if they lead to the construction of a new grammar of even greater generality or if they show some underlying principle is fallacious or misformulated."[24] That is, critics could not simply point out that something didn't work; they had to come up with a new theory in its place that did.

As relentless as the expansion of Chomsky's vision seemed to be, it deflated unexpectedly in the 1970s. Part of the appeal of generative linguistics was the way it rendered sentence analysis into mathematical-looking algorithms. Rules like "S goes to NP VP" gave language study a scientific veneer. It turned out, however, that this was not what actually happened in the brain as it processed language.

If deep structures really existed, it was reasoned at the time, you'd expect people to take longer to understand the more complicated, transformed structure of a given sentence than its simpler basic form. But when psycholinguists tested this in experiments, it did not pan out: the derived sentence took the same amount of time as the basic sentence.

Soon the voices that had criticized Chomskyan linguistics from the beginning grew to a din. As researchers found that the notion of an innate language organ was not supported by real-world evidence, they became interested instead in the idea of general foundations for language and thought. Even the popular press ran articles about the Chomskyan revolution and declared it over.[25]

For his part, Chomsky continued to dismiss objections to generative linguistics as being either uninteresting or not serious, and to assert that he had been misunderstood. And indeed, the history of modern linguistics is densely populated by straw men who look a lot like Noam Chomsky.

He was regularly accused of making statements that he had not. When he was charged with changing his mind or abandoning ideas that he once championed, he explained that he hadn't changed his mind but that he meant something else all along. While careful rereading of Chomsky's writing often bears out his claims, his great influence often worked against him. Chomsky's casual hunches and suppositions were often treated—and debated—as though he had made a fully defended argument.[26]

Certainly, Chomsky's terminology changed considerably over the years, and this must have contributed to his being misunderstood. In 1972 he referred to his developing ideas about language and the mind as the standard theory. In 1977 the standard theory became the extended standard theory, and later it became the revised extended standard theory. In the early 1980s Chomskyan linguistics was called principles and parameters theory, and then later government-binding theory. Over time, transformations were transformed into T-markers; phrase structure representations became P-markers. Instead of deep structure, surface structure, and logical form, linguists had D-structure, S-structure, and LF. Theta-theory described the assignment of roles like agent to noun phrases.

Some of the name changes marked big shifts in ideas. For example, in the earliest theories of UG, children were born with innate, very specific rules for languages. In the principles and parameters theory, children are born with a finite set of parameters for language that their experience of a particular language then modifies. So the differences in the syntax of different languages can be reduced to this collection of settings. Overall, though the many shifts make it hard to imagine that more than a few syntacticians can really track all the distinctions between them, a vision of language has remained consistent for all this time. Chomsky emphasized repeatedly both the complex nature of language and the fact that the human brain was especially designed to acquire and to implement it. As he wrote in 1975: "A human language is a system of remarkable complexity. To come to know a human language would be an extraordinary achievement for a creature not specifically designed to accomplish this task. A normal child acquires this knowledge on relatively slight exposure and without specific training. He can then quite effortlessly make use of an intricate structure of specific rules and guiding principles to convey his thoughts and feelings to others, arousing in them novel ideas and subtle perceptions and judgments.[27]

Declaring the revolution over turned out to be premature, and the downturn in the fortunes of generative linguistics was merely a blip. Just a few years after Chomskyan linguistics was supposed to be over, barely anyone remembered that it had been in peril. People continued to wax superlative at the mention of Chomsky's name, and comparisons to the great men of intellectual history kept rolling out: He was the Newton, the Einstein, of language. He was an intellectual colossus, a special kind of genius that made the merely normal geniuses look dim-witted. Not only did Chomsky's influence reassert itself, but in 1980 Charles Hockett complained of his "eclipsing stance." By now people didn't just think Chomsky's ideas were the most important thing in linguistics; they had begun to believe that nothing important had ever happened before Chomsky.

Writing about the many problems for Chomskyan theory in the 1980s that were simply ignored, the linguist and historian Peter Matthews likened the advance of generative linguistics in that period to the German army's march across France in World War II. (After World War I, the French built a huge fortification on the French-German border called the Maginot Line. When the Germans invaded France in World War II, they basically went around the fortification by going through Belgium, and from there they entered France unimpeded.)[28] Students continued to be attracted to Chomsky's work. One way of measuring the power of an academic is to count his intellectual children, the students he influences who leave the university, get jobs on other campuses and in other countries, and continue to teach the ideas of the teacher. These students' students become teachers and in turn influence their students. In this way, an academic lineage is created. Chomsky has been a prolific father; his heirs have gone forth and multiplied. The 1988 four-volume Cambridge survey of linguistics describes, for the most part, Chomskyan linguistics.

Says Steven Pinker, "The bulk of modern linguistic work has dealt with problems or phenomena that Chomsky noted." Still, even though Chomsky has had a powerful influence on other sciences, they have had a notorious lack of influence on him. All theories of language evolution in the last decade, as well as most ideas about language and the brain, are usually characterized as for or against him.

It's ironic that Chomsky, who began his career striking a blow against totalitarian ideas in the form of Skinner and who also happens to be one of the best-known radical-left figures in politics, is now himself a figure of totemic power. For decades, his name appeared in the synopses of conferences, the papers of students, and the articles of academics with all the frequency and duty that portraits of the leader appear in the classrooms of third-world dictatorships.[29]

How does one man inspire both blistering rage and religious devotion? There is little evidence to suggest that Chomsky has sought to create the sociological marvel that is his career. Academics who are familiar with him will—without exception—describe the way he insists that he is a minor figure with little real influence.

It is Chomsky's legend rather than any rationale that he advanced that stifled language evolution research during the latter half of the twentieth century. His public comments on the topic have mostly been cryptic. In his book *Language and Mind* he wrote, "It is perfectly safe to attribute this development [of innate mental structure] to 'natural selection,' so long as we realize that there is no substance to this assertion, that it amounts to nothing more than a belief that there is some naturalistic explanation for these phenomena."[30]

In the same book, Chomsky went on to wonder how many possible alternatives to transformational, generative grammar exist for an animal that evolved in the way humans did. Perhaps none exist, or only a few. If this were the case, he said, "talk about the evolution of language capacity is beside the point."[31]

In the 1980s Chomsky acknowledged that language must have given us some kind of evolutionary advantage but its origins were more likely to have been accidental than the result of slow evolutionary change. "We have no idea, at present," he said, "how physical laws apply when neurons are placed in an object the size of a basketball, under the special conditions that arose during human evolution."[32]

Certainly no one knew whether language was a function more of physics than of behavior or biology. Instead of resulting from adaptation and selection, language may have arisen as a by-product of a very complex mental machine. But at the time, few people engaged in any meaningful way with the idea. As a result, when confronted with this kind of Chomskyan koan, almost no one took the question of adaptation any further.[33]

Having stripped away all of the untidy bits of language as "performance," Chomsky defined language as an idealized, perfect, and elegant system. The brain, on the other hand, he said, was messy. How did something so messy develop something so perfect? It was a mystery, he said, one that was, for the time being, insoluble.

If it were true that language was perfect and that it simply emerged from our highly complex mental organization, Chomsky has also said, such a development does not make much sense with what we know about physical systems. Biology just doesn't work like that. Indeed, biological evolution is a haphazard, junkyard kind of process where traits are not intelligently designed from scratch, but rather, new tools are built over old ones. This conundrum was, in Chomsky's view, a problem for biology, not for linguistics. "What followed in theories of language acquisition," said A. Charles Catania, a professor of psychology at the University of Maryland, Baltimore County, "was closer to creationism than any other part of psychological research."

So, while Chomsky did publicly discuss the utility of language, whenever he mentioned evolutionary theory, it was mostly to discourage its value as a solution to the origins of language. He said, reasonably enough, that you can't assume that all traits are selected for. In one of his most concrete statements on the topic, he wondered aloud whether a genetic mutation might have been responsible for the property of discrete infinity, which he considered fundamental to language.

As far back as 1973 critics had complained that "the notion advanced by Chomsky among others, that a language system could have come into existence suddenly, as the result of a 'mutation,' seems simplistic and hardly more plausible than the idea that language is a gift of the gods."[34] Yet Chomsky in no sense advanced this argument; he merely suggested it. His most damning evaluation of the idea that language was an adaptation was that it was "hard to imagine a course of selection that could have resulted in language."

Such was his eminence that when Chomsky said things like it's "hard to imagine," it was taken to be a truth about the intractable nature of the problem rather than the limits of imagination. It is a testament to his rhetorical skills and the depth of his influence that a strong case could be so widely inferred from his highly qualified statements on the topic.

Against the backdrop of Chomsky's rather pointed lack of interest, the problem of language evolution remained for most of the twentieth century the domain of the occasional crackpot and a few brilliant and determined mavericks. Sue Savage-Rumbaugh belongs to the second group. While the consensus in linguistics and most of psychology was that language was a monolithic trait that only humans possessed, Sue Savage-Rumbaugh was busy trying to teach another species how to use it.

2. Sue Savage-Rumbaugh

I t's no exaggeration to say that Chomsky entered the academic scene with a crash, announcing his interests in such a compelling way that generations of scholars fell into lockstep with him. Yet despite his dominance, islands of research have sprung up independent of his school of thought. For the last few decades, ape language research has been one such island.

Social, affectionate, emotional, and smart, apes need other apes, just as humans need other humans. This seems obvious enough in the twenty-first century, but it is relatively recent knowledge, the fruit of painstaking observation by primatologists like Jane Goodall.[1] The notion that human intelligence was a unique phenomenon started to break down in a very small way with the birth of primatology. The field's findings have become so ingrained in popular consciousness that it's now very hard to believe that as recently as fifty years ago we knew virtually nothing about apes and other primates. The years that Goodall and her colleagues spent patiently watching them in the wild yielded powerful insights, not just into the lives of other primates but also into how like them we are.

Robert Sapolsky, a longtime observer of baboons (which are in the monkey family and therefore more distantly related to humans than apes are), draws attention to the similarity of our emotional and cognitive lives in his description of a mother baboon's mishap:

One day, as she leapt from one branch to another in a tree with the kid in that precarious position, he lost his grip and dropped ten feet to the ground. We various primates observing proved our close kinship, proved how we probably utilized the exact same number of synapses in our brains in watching and responding to this event, by doing exactly the same thing in unison. Five female baboons in the tree and this one human all gasped as one. And then fell silent, eyes trained on the kid. A moment passed, he righted himself, looked up in the tree at his mother, and then scampered off after some nearby friends. And as a chorus, we all started clucking to each other in relief.[2]

The intelligence, the shared attention, and the intense sociability that Sapolsky noted cannot help but remind us of our own species. Such similarities, according to Darwin, were likely inherited from a common ancestor. Indeed, he argued that the traits we have in common with a closely related species are a matter of shared inheritance rather than independent, parallel evolution. So if we want to look at early stages of linguistic development, it makes sense to examine our tree-dwelling and generally less-inhibited cousins.

Sue Savage-Rumbaugh's name may not be as familiar as Noam Chomsky's, but her place in history is assured. She is the researcher who has most successfully bridged the species gap by teaching an ape to produce and understand aspects of language. She and her colleague Duane Rumbaugh take raw material like a chimpanzee or bonobo, with its familiar neural architecture, and see to what extent they can bypass a few million years of evolution.

Before Savage-Rumbaugh began work with Kanzi, a bonobo, other ape studies had successfully taught chimpanzees to comprehend language. The problem was, as Savage-Rumbaugh pointed out, that even though creatures like Washoe could successfully use language to request food or obtain other objects of desire, they weren't any good at taking on the other role in the communication process. For Washoe, Sarah, and Lana, the first generation of language-trained apes, wrote Savage-Rumbaugh, language was a one-way street. It only functioned as a tool for getting what they wanted; there was no listening.

One of the first and most important discoveries for ape language research (ALR) was that trying to teach language directly was not the way to go about it. ALR, which began in the 1970s, made an evolutionary leap when Savage-Rumbaugh realized that apes were best taught indirectly rather than explicitly. Savage-Rumbaugh had been trying to teach language to Kanzi's mother, Matata, for a number of years. During this time, Kanzi had simply observed the two in their lessons. On the first day that Savage-Rumbaugh turned her attention specifically to Kanzi, he spontaneously used the picture keyboard to combine symbols and communicate to her what he wanted her to do and what he wanted to do next. Kanzi had been learning language all along. "I was in a state of disbelief," wrote Savage-Rumbaugh.[3] (The same process applies for human children. Even though they typically receive some explicit instruction, such as leafing through a picture book with a parent and associating animals with their names, children primarily acquire language by hearing it around them and by interacting with creatures who speak.)

Thereafter, instead of being formally instructed in the value and use of a language system (imagine trying to introduce the concept of verb tense to a classroom of apes), the bonobos were raised in a language-rich environment. While Washoe had never learned a sign without being taught with hundreds and hundreds of repetitions, Kanzi, and soon another bonobo called Panbanisha, picked up words by being regularly spoken to during feeding, playing, and grooming; having symbols on the picture keyboard pointed out to them with the spoken word; and even watching television. Such activities were all that was required to outfit Kanzi and Panbanisha with some language skills.

Over many years, these two apes learned how to manipulate keyboards that contained visual images, of milk or a dog, say, instead of letters. They also learned how to comprehend spoken English, coming to understand hundreds of single words and longer constructions. (Unlike other experiments in which monkeys perform for food rewards, these apes have free access to food all day.) Kanzi and Panbanisha are able to participate in two-, three-, and four-way conversations. They can converse about objects as well as intentions and actions, and state of mind. Testing has shown that Kanzi in particular is capable of correctly understanding hundreds of sentences that he's never heard before, sentences like "Show me the ball," "Get me the snake picture," and "Can I tickle your butt?"

As well as developing comprehension abilities at the level of a three-to-four-year-old child, the bonobos demonstrate creativity in their manipulation of language. They spontaneously combine single words they already know to create new words, like linking "water" and "bird" as "waterbird" to mean a duck. They've also been known to make up sentences in response to novel situations. The ape Sherman, who was raised in a different experiment, once rushed into his lab in order to tell the scientists inside, "Scare outdoors." Sherman had just seen a partially anesthetized ape being carried past in a stretcher.

Still, sometimes even the cleverest primates have difficulty with comprehension. At the March 2002 Evolution of Language conference at Harvard, Heidi Lyn, who was working at the time in the Language Research Center at Georgia State University, recounted what happened the day that Savage-Rumbaugh told Kanzi to put water on a carrot. The ape threw the carrot outdoors. Thinking he had misunderstood, Savage-Rumbaugh repeated the request. In response, Kanzi pointed vigorously outside. It was raining.

Lyn is now at the University of St. Andrews in Scotland, where she is involved in a dolphin research project. She is also writing a book that brings together the findings from all of the animal language studies. She has worked with Kanzi, with language-trained dolphins under Lou Herman in Hawaii, and with Diana Reiss in New York on a dolphin keyboard project. The earliest animal language experiments, Lyn explained in an interview, began in the 1890s, with documented cases of people raising apes in human homes, and in some instances raising them side by side with human children. It wasn't until the 1960s through the late 1970s, however, that scientific animal language research really boomed.

The early attempts to get apes to communicate like humans were failures, primarily because researchers were trying to induce apes to talk. This focus changed when Allan and Beatrix Gardner, a husband-and-wife team at the University of Nevada in Reno, perceived that apes seemed to find gesture easier than vocal communication. The Gardners reared Washoe, a female chimpanzee, in their home, teaching her a modified version of American Sign Language. Washoe was extremely successful and learned hundreds of different symbols. She was rigorously tested again and again, and her learning stood up. In 1972 Penny Patterson, a Stanford Ph.D. in developmental psychology, began her lifelong experiments teaching sign to Koko the gorilla. Duane Rumbaugh also began to work with the chimpanzees Lana, Sherman, and Austin, seeing how well they could communicate with picture symbols, called lexigrams. There was enormous interest in this work and many interesting results, said Lyn.

In the 1970s a young academic named Herb Terrace heard about the Washoe work. He was excited by the results and wanted to replicate them, so he obtained a chimpanzee and called him Nim Chimpsky. Terrace followed the Gardners' work closely, although he had many more people interact with Nim than had ever interacted with Washoe. Initially, it looked as if he had successfully taught Nim to use words and some syntax. But when he did a frame-by-frame video analysis, he realized that what Nim was doing was less symbolic than imitative: Nim wasn't using language independently but instead responding to cues that Terrace or other caretakers were giving him. At the same time Terrace also did a video analysis of Washoe and Koko and concluded they, too, were being inadvertently cued by their handlers and neither thinking nor communicating. He published the results of his investigation in the journal *Science* in 1979.

The damage from Terrace's findings was immediate and devastating. His article was picked up by the press, and a popular and scientific consensus quickly developed that the apes weren't doing anything their caretakers hadn't cued them to do. Funding for animal language research very rapidly dried up. The Gardners were effectively shut down, and one of their graduate students, Roger Fouts, took over and was for a long time only able to maintain but not expand the Washoe project.

From that point on, said Lyn, it became very hard to get any animal language data published. After the Nim Chimpsky publication, Lou Herman started his studies with the dolphins Akeakamai and Phoenix, using an artificial language and focusing on comprehension (his funding for the project was secured before 1979). The fact that he was concerned with comprehension, rather than production of language, was probably what saved his work, said Lyn. People found it easier to consider the possibility of animals' understanding versus producing language. Still, Herman didn't publish his first paper until 1984. As soon as the paper came out, he was criticized intensely for using linguistic terms like "sentence," and "noun," and "verb" to describe what the dolphins were doing. That response was unjustified, said Lyn. In fact, she said, Herman has the best data on syntax for any animal, anywhere. Akeakamai and Phoenix have mastered a complex grammatical system. If Herman gives the dolphins nongrammatical sentences, they will either refuse them or make grammatical sentences out of them.

A year after Terrace's *Science* article was published, Martin Gardner reviewed a number of books about animal language training in the *New York Review of Books*. He began by tracing a direct line from crackpot claims that dolphins communicated through ESP to ape language research. His first pass at evaluating Penny Patterson's work with Koko and the attention it received had more to do with Patterson herself than with her science. "It is not hard to understand why Penny—young, pretty, with long blond hair—has received such enormous publicity," he wrote. "What could be more dramatic than color photographs of Beauty and the Beast, heads together, raptly chattering to one another?"

It is hard to understand how comments like Gardner's become part of the debate: the same would never have happened had the scholar in question been, say, Chomsky, who has likely never had his physical appearance assessed in reference to his work and its public appeal or been called "Noam" in similar circumstances.

Apes might have a "feeble talent" for putting together signs in meaningful ways, but it was more likely, Gardner concluded, that ape language research amounted to little more than an unconscious collusion between a cooperative animal and a hopeful human. As he wrote: "There is no solid evidence that an ape has ever invented a composite sign by understanding its parts. In the course of several years an ape will put together signs in thousands of random ways. It would be surprising if it did not frequently hit on happy combinations that would elicit an immediate Clever Hans response." (Clever Hans was a famous horse who could allegedly perform mathematical computation. He would indicate the answer to a problem by pawing at the ground the correct number of times. A 1907 study showed that Hans's owner gave him subtle and unconscious cues when to stop pawing at the ground.)[4]

Terrace did make some important contributions, explained Lyn, by pointing out that there had been no scientific controls in the studies assessing the apes' syntactic ability. Mostly, the claims for syntax were based on naturalistic observations and had not been rigorously tested. But because Terrace found instances of cuing, the scientific community and the public decided that all of the behavior was cued. There were, in fact, numerous examples of solid, double-blind experiments, such as one where Washoe was placed alone in a room. A camera was trained on her, and pictures were flashed up on a screen before her. The chimpanzee made the signs for every object in the pictures, and because she was by herself, cuing was impossible.

Luckily for Savage-Rumbaugh, her funding had been renewed for five years just before the Terrace article appeared. She spent those years producing valuable findings. For example, Kanzi and Panbanisha have spent time with other apes in different experimental situations. For a while, they were raised with another bonobo, Tamuli. But while Kanzi and Panbanisha were exposed to language from the time they were just a few weeks old, Tamuli's exposure began much later in life. She was initially reared by her mother, but at three and a half years of age she was allowed to accompany Kanzi and Panbanisha in their daily activities, like taking trips to the forest. Kanzi and Panbanisha's human caretakers also spoke to Tamuli while pointing at the picture keyboard and describing their daily activities.

Tamuli never developed language skills comparable to those of the other apes. In this respect she is like human children who, for whatever reason (for example, undiagnosed deafness or abuse), are not exposed to language at an early age. There is a crucial learning period for humans when they must be exposed to language. Even if they are neurologically normal, they will never fully acquire language if its foundations are not laid in this early period of brain development. Genie, the most famous of these cases, was kept locked in a room, denied normal human communication, and never taught language, and by the time she was rescued, she was unable to acquire much more than basic language skills. Her experience shows that if you are denied language, you don't spontaneously produce it. Since Genie's case was studied, it has become fairly well established that language is not innate in the same way as, say, our instinct to breathe or cry.

Tamuli's experience suggests that apes have a similar window of opportunity.

Even though Tamuli could not relate at the sophisticated level that Kanzi and Panbanisha did, she at least seemed to understand that the keyboard was intended for communication. While the apes often tried to use it to relay messages, Tamuli's usage was a bit like that of a young child banging away on a piano or a keyboard. She made no sense.

One thing that Tamuli lacked was the ability to recognize that her interlocutors had separate minds and that communication with them could alter their perceptions. Kanzi and Panbanisha, on the other hand, seemed to have acquired a theory of mind along with language. When Panbanisha saw her trainer remove candy from a box, replace it with a bug, and then give the box—supposedly still with candy—to Kanzi, she called her "bad." The chimp demonstrated that she could understand what was going on in her trainer's mind independent of the reality and apply language to the situation. She herself scared Kanzi when she used language to tell him there was a snake nearby, when in fact there was no snake. Panbanisha used language to manipulate the contents of Kanzi's mind, just as her trainer had manipulated the contents of the box.

The ape experiments indicate how memory is a vital component of language use, even at this rudimentary level. A chimpanzee, Panpanzee, who was raised with Kanzi and Panbanisha, would sometimes make language mistakes that demonstrated a limited memory, such as when she was asked to put a sweet potato in the microwave. Typically in an experiment like this, an object, like the potato, would be made readily available for Panpanzee, placed right before her eyes. But in this example, the chimpanzee had to retrieve one from the refrigerator in order to complete the request. This she did. But instead of putting the vegetable in the microwave, she took it to the sink and proceeded to wash it. Somewhere between the retrieval process and the end task, the request became scrambled for the chimp. In similar situations, Panpanzee's incorrect response suggested she was falling back on her knowledge of routines, rather than correctly remembering a novel request (something people occasionally find themselves doing as well).

Sometimes language mistakes can be as useful as correct responses.

Eliciting errors in human speech is one of the main methods that psycholinguists use to expose the mental strategies that underpin language use. Spoonerisms, for example, aren't just sound swaps: "pea tot" (teapot), "dood gog" (good dog), and "band hag" (handbag) suggest that speech is not entirely spontaneous. If a speaker accidentally begins a word with the first sound of the next word, he must be planning what he is about to say, even if he is not aware of it.

Lyn analyzed eleven years' worth of Kanzi's and Panbanisha's language error data and found that when the apes accidentally pressed one keyboard picture instead of another, or when they misunderstood a spoken word, their errors usually revealed an underlying connection between the intended word and the mistaken one. Just like humans, the apes made category substitutions, like mistaking colors, such as red for black. They made word association errors, confusing the names of locations with items that were found in those locations. And they made phonological (sound) errors, like using a word because it rhymed with the intended word.

Lyn and colleagues found that Panbanisha and Panpanzee have more symbol ordering rules in common with each other than with their caretaker. It's possible, even probable, that the last common ancestor between these apes and humans had the ability to understand meaning-based ordering strategies. Lyn also found that these apes have a gesture-last rule: they always touch the lexigram, and then gesture in the real world.

Bonobos acquire language up to the level of human children. For example, they can understand sentences that contain one verb and a three-noun phrase ("Will you carry the M&M's to the middle test room?"), but they have trouble with conjoined sentences that require two separate actions ("Bring me the ball and the orange"). They do not speak English words, though they attempt to do so. Their short-term memory seems to be only half the capacity of human children's, so they are not as good at imitating a series of utterances without a lot of repetition.[5] The more complicated syntax gets, the more trouble they have with it.

The ape language research led Savage-Rumbaugh and her colleagues to conclude that language consists of "a large number of component parts and interacting functions."[6] Even though their work has not had the impact of Chomsky's, most researchers in language evolution would today think about language in these terms.

What's most striking about the older criticism of ape language research is its basic attitude, which is more motivated to discredit than evaluate. In much ALR commentary, there is a strong sense that the critics have already made up their minds before arguing or offering reasons why ape language couldn't work. There are claims of falsifying data and even of people being out to get each other. In the 1980s the debate was rarely conducted without tones of disdain and contempt.

Even now, scholars who work with animal language are often characterized as daft idealists or outright frauds, believing that beneath the fur or behind the beak are creatures with souls. Yet if you speak to these researchers, you won't find anyone downplaying the enormous differences between humans and other animals, despite the fact that they happen to be interested in the commonalities.

One legacy of the Terrace paper has been an ongoing difficulty getting funding for this kind of work. Researchers often have to go outside the typical funding bodies of academia to keep their studies going, turning to special interest groups and private individuals. The promotional literature for the Koko research mentions visits from Sting and Robin Williams, for example, a gambit that gives animal language research a weird profile. Such marketing gives the impression that it is not solid, straightforward science.

Still, the basic tenor of the commentary has begun to shift over the years. Critics used to dismiss the research by saying, "All that the animals have is a few words, and they don't have any syntax whatsoever." Now the fact that apes can acquire words is treated as an interesting phenomenon.

Frans de Waal, a professor of primate behavior at Emory University and the author of *The Ape and the Sushi Master,* says:

> I think the trend is clearly towards poking holes in the wall that exists between us and animals, and increasingly people embrace the comparison, so to speak. In the 1970s, when I had to give a lecture on

chimpanzees, some people would say, "How can you use the term 'rec-
onciliation'?" They would have strong objections. Or let's say it was
about sex differences, and they would say, "How can you compare
chimpanzees and humans?" Because obviously, we are cultural beings
and we can change our behavior.

When I give lectures on these topics today, that never happens anymore. It's because there's a gene on the cover of *Time* or *Newsweek* almost every week, a gene for this or a gene for that, so people are
getting very used to the idea that genes add something to behavior. So
the climate is totally different, and there's a much greater openness to
seeing us as animals, as Darwin always wanted and as many other people wanted.

I was recently invited to give a talk for business ethicists. Now,
business ethicists are basically philosophers who teach at business
schools. Even there, there is an enormous openness for these comparisons, whereas I'm sure twenty years ago they would not want to even
touch a monkey. So I think the trend is clearly towards more comparisons. More comparisons doesn't necessarily mean that we fully accept
the similarities. Usually they'll want to keep something like, "This is
typically human" or "This is unique to humans"—they want to keep
this to some degree.

One of the most important contributions of ape language research
is its challenge to the traditional idea that other animals have a fixed mental bag of tricks, and humans are different because we have language and
that makes us mentally flexible. If that were the case, Kanzi would have
been unable to learn the language skills he has. Clearly, these apes who
have the rudiments of language can also be flexible and creative with their
communication.

Ape language research, and Kanzi in particular, opened one fascinating
window into the problems of language evolution. Steven Pinker and Paul
Bloom opened another in 1990 when they published a paper in which they
sidestepped the question of how much animal language training can teach
us about language evolution and instead argued directly that not only could
language evolution be studied but it *should* be studied. The two scholars—

one a rising academic star, the other a graduate student with a brilliant idea—inflamed hearts and minds because their proposal was clever, innovative, and engaging. And even though they weren't the first to propose that language evolution was a valid topic of inquiry, their paper ignited a small blaze that quickly grew and spread.

3. Steven Pinker and Paul Bloom

In 1989 Paul Bloom, a twenty-five-year-old graduate student in the psychology department at MIT, was doing research in child language development. He was interested in word learning in young children, which had nothing to do with evolution, but he was increasingly bothered by the general agreement that language could not have evolved.

"Two things happened at once," he recalled in an interview.

One was that Leda Cosmides, who is now at the University of California, Santa Barbara, came to give a talk at MIT. She's a prominent evolutionary psychologist, and she started talking about the mind and language from an adaptive point of view. When we met later, I said, "This is ridiculous!"

I responded to her with the Stephen Jay Gould line, which I had totally been persuaded by years before. There was no reason to favor an adaptionist account of language (as opposed to the view that it was an evolutionary accident).

She was very civil and intelligent, and she said, "No, no, you're mistaken." And she convinced me that it made sense to apply an evolutionary analysis to mental life. Some things may be artifacts of biology, but there are good reasons to believe that something as rich and as complicated as language could have evolved by natural selection.

It was one of the rare case where an academic changes his mind. After thinking about it for a while, I realized that it made sense.

And then, at the same time, Massimo Piatelli-Palmarini, a colleague and friend of mine in the Department of Linguistics and Philosophy, published an article in *Cognition* on the evolution of cognition and language. His article presented in this very sharp, cogent fashion the Chomskyan view on evolution—basically he said that there was very little interesting to make of the connection between natural selection and cognition and that language has features that simply cannot be explained in terms of adaptation. I strongly disagreed with it.

At the time that the *Cognition* article appeared, it looked to Bloom as if everyone else agreed with Piatelli-Palmarini. "Back then if you didn't independently have an interest in evolutionary biology or evolutionary theory, the arguments of Chomsky, on the one hand, and Gould, on the other, were very persuasive. Chomsky is the smartest guy in the world and the dominant figure in linguistics, and Gould is this lay saint, this wonderful writer and brilliant synthesizer. And they're both telling you the same thing—that language didn't evolve as a result of natural selection.

"You can't underestimate the influence that Chomsky had," said Bloom. "People believed this line partly because of the force of Chomsky's personality. A linguistics friend of mine told me in all seriousness about what he called the C-principle. The idea is that if Chomsky believes something, then it makes sense to agree with him in the absence of other knowledge. Because, you know, he is a really smart guy.

"There was also something of an ideological taint about adaptionist explanations," remembered Bloom. "It was a sort of a dark association with racism and sexism and the evils of biological determinism, and people were wary of being associated with that.

"So I approached Steve Pinker." Bloom was acquainted with Pinker as a young professor in the psychology department who studied language.

I don't know whose idea it was originally, but we discussed writing a response to Massimo's article. We did not disagree with Massimo about his characterization of language. We did buy the Chomskyan party line that there was an innate, mental language organ. But we disagreed about evolution.

So I wrote up this little thing. It was five pages long, something like that. It was very drafty. And I gave it to Steve, and he came back to me with this thirty-page thing. It was monolithic and far more ambitious than the paper I had written. At that time neither of us knew much about evolutionary biology or the issues in detail, and so we were both reading up on it. Trying to keep up with Steve when he's acquiring new knowledge was a difficult task. At that point, it definitely became "Pinker and Bloom," not "Bloom and Pinker." Steve was the dominating intellectual force here.

Stephen Jay Gould, whose line Bloom had taken with Leda Cosmides, was at the time an intellectually flamboyant and highly influential evolutionary biologist. He was based at Harvard as the Alexander Agassiz Professor of Zoology, and he believed passionately in spreading the word about evolution. For years he wrote essays for *Natural History* magazine, many of which had been collected into popular books like *Bully for Brontosaurus* and *The Lying Stones of Marrakech*.

Gould's *Natural History* column was widely read within the academic community, and his books sold extremely well to both specialist and popular audiences. He wrote with enormous verve about the lessons and mysteries of evolution, ranging from the subtleties of natural selection, "the wriggles of a million little might-have-beens," to faked fossils, racism in science, and the most singular minds of scientific history.

The "Stephen Jay Gould line" was that scientists were too quick to apply evolutionary explanations to everything. Some features of our lives did not result from adaptation, he argued, but are just accidental by-products of other evolutionary changes. Gould called these biological artifacts "spandrels." As he explained:

Since organisms are complex and highly integrated entities, any adaptive change must automatically "throw off" a series of structural by-products—like the mold marks on an old bottle or, in the case of an architectural spandrel itself, the triangular space "left over" between a rounded arch and the rectangular frame of wall and ceiling. Such by-products may later be co-opted for useful purposes, but they didn't arise as adaptations. Reading and writing are now highly adaptive for humans, but the mental machinery for these crucial capacities must have originated as spandrels that were co-opted later, for the brain reached its current size and conformation tens of thousands of years before any human invented reading or writing.[1]

Throughout his career, Gould stressed the ways in which the human species was a glorious accident. The wonder of evolution, he emphasized over and over, was that it was "an unpredictable process with no drive to complexity."[2] In life, there is only forward motion, just the drive to keep driving. At some point in the past, Gould believed, our brains evolved to a level of complexity that would enable us to reason our way through certain situations, and at that level we had the structures for language already in place. In a sense, language simply "happens" when you have a machine complex enough to accommodate it. So rather than language being selected, we lucked into it, and it wasn't part of what initially made us successful as a species—even though now it's essential to our existence.

In 1997 Gould gave a talk at Iowa State University. It was one of probably hundreds he presented as one of the century's most ardent popularizers of evolutionary theory. And it went, no doubt, as most of those talks did. Gould, short and remarkably loud, spoke with vigor about evolution. After his speech, he spent a lot of time answering questions about evolution and equal amounts of time batting away the creationists who had come to bait him. When someone asked about the evolution of language, he was uninterested, even a little annoyed by the question. He waved his hands about and said, "It's probably a spandrel."

Steven Pinker was thirty-five years old in 1990. A decade earlier he had completed his Ph.D. thesis in an unusually short amount of time. He was hired by Harvard in 1980, and was lured to Stanford in 1981, only to be lured to MIT in 1982. Pinker began to work there on regular and past-tense forms of verbs and how children acquire them. When Paul Bloom approached him, he had not thought a lot about evolution, but he eagerly dove into the research. "I was motivated," he said, "by the feeling that there was a premature consensus from two charismatic figures who did not have a sensible argument.

"On the one hand, there was the Gould-inspired consensus that we were questioning. And the thing about Gould was that his views were not mainstream within evolutionary biology though people outside the field were not aware of that. And there was also the Chomsky viewpoint." He continued:

> It's by no means the case that everyone in child language acquisition or cognitive science in general is a Chomskyan. He's a deeply divisive figure. But there are large sectors that are in almost religious thrall to him. If he says it, it must be true, and if you disagree with him, then you must misunderstand. Non-biologists even get their evolutionary biology from Chomsky's footnotes. I remember Chomsky made a throwaway mention in a footnote of an argument by a mathematician and an engineer that natural selection could not work. It was a back-of-the-envelope calculation, whose flaws were immediately pointed out by biologists, and no one but Chomsky ever took it seriously again.

Pinker is now back at Harvard. His suite on the ninth floor of William James Hall is airy, spacious, and clean. The walls are lined with books, and a large table with room for six, as well as space in the middle for one-on-one discussions. Against the wall near a large window is Pinker's desk, and on it a brass statue, the Emperor Has No Clothes Award, from the Freedom from Religion Foundation. At the other end of the room, behind a sliding whiteboard, is a brain in a jar. Pinker himself, apart from the famous flop of curls seen in his many author photos, is contained, his comments brief and well measured.

As a first-year student Pinker had cross-registered for a course at MIT taught by Noam Chomsky and Jerry Fodor. When he became a professor at the school, he attended a few lectures Chomsky gave, but he never worked directly with him. "Although in the grand scheme of things I'm probably closer to Chomsky than many people in cognitive science," he said, "I'm not part of the cult of personality that has grown around him." Pinker is now the Johnstone Family Professor of Psychology at Harvard. In response to a question about why people were so willing to take one of the most fundamental questions about language—how it evolved—on faith, he replied, "There were a few reasons. Chomsky was a well-known left politician, and people perceived sociobiology, as it was then called, as right-wing. Truly, it doesn't need to be seen that way.

"Also, academics are lazy. They are unwilling to make their discipline rigorous in terms of the standards of another discipline, and that's how it was with evolution and cognitive science for a long time."

After a few months Pinker and Bloom wrote their paper up for the MIT Occasional Papers series. These technical reports are circulated throughout the university and sent to interested individuals outside it as an opportunity for researchers to get commentary from their peers.[3] In it they wrote: "Noam Chomsky, the world's best known linguist, and Stephen Jay Gould, the world's best known evolutionary theorist, have repeatedly suggested language may not be the product of natural selection but a side-effect of other evolutionary forces such as an increase in overall brain size and constraints of as yet unknown laws of structure and growth." As a result, they said, "in many discussions with cognitive scientists, we have found that adaptation and natural selection have become dirty words."[4]

Pinker and Bloom continued with an appeal to rationality. "In one sense our goal is terribly boring," they wrote. "All we argue is that language is no different from other complex abilities, such as echolocation or stereopsis [the visual process that gives rise to depth perception], and that the only way to explain the origin of such abilities is through the theory of natural selection."[5]

An e-mail exchange between the two authors and Chomsky ensued. In his e-mail, Chomsky made a series of unambiguously clear statements about the evolution of language. He said that he was not at all opposed to the idea that language evolved—of course it did—and that many parts of it were adaptive for communication. But he had great reservations about whether what he and serious linguists called language—the unique mental syntactic component—originated in the act of communication. He reiterated that there were factors in evolution other than natural selection, which were as likely to be significant. And in this regard, Chomsky, Pinker, and Bloom were essentially in agreement, their debate arising more from differing emphases than actual discord. Pinker and Bloom were still generativists at heart, and their goal was to discover where evolutionary theory and generative grammar were compatible. They also said that natural selection couldn't explain everything about the evolution of language. Yet they questioned how much "as yet undiscovered theorems of physics" would explain language's intricate design. In their e-mail to Chomsky, they wrote, "No matter what the constraints are on how you can grow a fin in a biological system, you need an explanation as to why fish have them and moles don't."

Certainly, the researchers also disagreed on fundamental issues, if not about what the key aspects of language were, then about how much they mattered in evolution. Though they concurred that language was indeed used for communication, they differed on how much this mattered for natural selection.

Ultimately, they disagreed most in what they felt the value of the debate was. On the one hand, Chomsky believed many of the relevant issues were either too trivial or too hard, and on the other, Pinker and Bloom claimed the study of language evolution was neither too mysterious nor too challenging to grapple with. It was, instead, a productive and scientifically valid endeavor.

Before they launched their argument about adaptation and natural selection, the authors reiterated some important and, at the time, well-accepted facts about language. For example, as far as we know, humanity has always had language. There were no creatures that we would think of as effectively human, no highly organized societies of people that hunted, gathered, and nurtured their offspring through a long period of vulnerable infancy, without language.

Additionally, pretty much everyone agrees that all languages are equally complex.[6] English, the dominant language of the United States with its advanced technology, is no more or less complicated than the language of the Andaman islanders off the coast of India. Moreover, said Pinker and Bloom, Modern English is no more complex than the English of six hundred years ago. Anyone who's tried to read Chaucer knows that Middle English is painfully different from today's English, but even though it has undergone enormous change, our language is in no sense an improvement on Chaucer's.

Children, said Pinker and Bloom, master complicated grammars by the age of three without any formal instruction. They can make grammatical distinctions that no one has ever demonstrated for them. Once they have acquired language as adults, brain damage can severely affect their language but leave other mental abilities intact. Or it can happen the other way around, with rich and complex language skills continuing to exist in a brain that has trouble with other, simple tasks.

They particularly emphasized that language is incredibly complex, as Chomsky had been saying for decades. Indeed, it was the enormous complexity of language that made it hard to imagine not merely how it had evolved but that it had evolved at all.

But, continued Pinker and Bloom, complexity is not a problem for evolution. Consider the eye. The little organ is composed of many specialized parts, each delicately calibrated to perform its role in conjunction with the others. It includes the cornea, the transparent, dome-shaped tissue that covers the front of the eye and refracts light, and the colored iris, which, like the aperture of a camera, controls the amount of light that enters the eye. Exceedingly tiny muscles pull apart the opening of the iris—the pupil—or shut it down, depending on the amount of light hitting the eye. The lens, suspended by fine fibers behind the iris, adjusts its own shape, so that the eye can focus on objects that are very near or very far. And the retina, layers and layers of differently specialized tissue at the back of the eye, takes the light entering the eye and translates it into a biological signal that's transmitted along the optic nerve to the brain.

All of these tiny, perfect biological devices operate in brilliant conjunction with one another to produce vision. The paradox of the eye is that evolution occurs in extremely small steps, yet it makes no sense for an eye to have evolved piece by piece. A cornea wouldn't begin to grow without the rest of the eye around it, and the same goes for all the other components. What about the vitreous humor, the goo that plumps the small globe up? Did it arrive with a sudden squirt, or did it inflate the eye slowly over time? It's infinitely unlikely that some lucky creature was one day born from unseeing parents with a complete eye in its head. Even Darwin said that it was hard to imagine how the eye could have evolved.[7]

And yet, he explained, it did evolve, and the only possible way is through natural selection—the inestimable back-and-forth of random genetic mutation with small effects and then the selection creatures with those effects by nature. It evolved to meet a specific need—vision. In the case of eyes, each time a small random change increased a creature's ability to register signals from its environment, that ability meant that it was likelier to survive and have offspring, and then its progeny got to pass on those changed genes. Over the eons, those small changes accreted and eventually resulted in the eye as we know it.

In the same way that the eye evolved to meet the need of seeing, said Pinker and Bloom, language evolved to meet the need of communication. The survival advantages of our kind of communication are as obvious and profound as the survival advantages of our kind of vision. Language enables us to learn from others. Humans don't have to experience something directly to know that it's either a good or a bad thing to do. If we have been warned about them beforehand, we can stay away from dangerous situations and move toward safer ones. The more we all get to share with one another, the more collectively we know.

Because this talking network is so important, knowing what our interlocutors are feeling, thinking, and meaning is also pretty important to survival, and language is superb for interpreting the thoughts and feelings of others as well. Moreover, there are distinct advantages to evolving a language that uses sound as a medium. If you're talking rather than, for example, signing, you don't have to look at someone, you don't even have to see him or be seen by him; it could be the dark of night, you could be hiding behind a tree, and your hands and body would remain free to do other things.

The kind of communication we specialize in, said Pinker and Bloom, is the production of propositions: "I am hungry"; "There's a bear"; "You are cute." And the communication of propositions is fundamentally connected to the channel in which it occurs—sound, from mouth to ear. This means that propositions occur one after the other, not all at once. Language is essentially serial.

The building blocks of serial communication, they explained, are nouns and verbs and the rules of structure and sound with which we put them together. They allow us to talk about events, objects, places and times, agents and patients, our intentions and others'. Words and rules allow us to build complicated sentences from smaller ones, and they help us pick the right meaning in an ambiguous statement.

Pinker and Bloom stressed again and again that even though what they were suggesting would be novel for cognitive scientists, it was not new for evolutionary biologists. All we are doing, they insisted, is applying the same kind of reasoning biologists apply when they discover complicated systems in other animals. ⤙

Pinker and Bloom originally planned to send their paper to *Cognition* as a reply to the Piatelli-Palmarini article. "But very quickly it grew," said Bloom, "and we decided to send it out as a freestanding article to *Behavioral and Brain Sciences*."

For Bloom, working with Steven Pinker was a thrill. "He was always understood to be a genius," said Bloom.

> He has a reputation as a genius. But while there are a lot of smart people who are full of themselves and difficult to work with, Steve is a mensch—very intellectually generous and kind.
>
> We submitted the paper, then another thing happened. Steve and I were asked to give a talk at the MIT Center for Cognitive Science seminar series. And so we were set to give this talk and expound the position of this paper, and then I found out that the commentators would be Stephen Jay Gould and Noam Chomsky.
>
> I was absolutely terrified. Besides his obvious status in the field— he's like the Descartes of our time, people will look back a thousand years from now and will know his name—Chomsky is utterly merciless in debate, and I didn't really want the experience of getting my ass kicked by him.
>
> And there were other people who were very unhappy about the article. One colleague of Steve's at MIT was extremely upset. He was very much of a Chomskyan and was really mad. He thought we were being hugely naive about evolution and wrote a long letter that was quite angry, accusing us of all sorts of things.
>
> And I think a lot of people were really upset in part just because we disagreed with Chomsky. A lot of that anger was directed at Steve. I

might have been thought of as a poor graduate student who was se-
duced into it, an Oliver Twist to Steve's Fagin. But Steve was at MIT,
Chomsky was at MIT, and I think some people felt it was betrayal.
You'd expect it from Phil Lieberman at Brown or Elizabeth Bates at the
University of California, San Diego, they were always disagreeing with
Chomsky, but Steve was at the center of things.

That evening, something happened. I think Chomsky's back went
out and he couldn't do it. I felt *transcendent* relief. Chomsky was later
replaced by Massimo Piatelli-Palmarini.

Still, going up against Stephen Jay Gould in debate was no small feat for
any academic, let alone a student:

We met before for an informal dinner, but I was too nervous to eat.
When we got there, the auditorium was packed, and they sure as hell
weren't there to see me.

The room was crowded for Gould, but a lot of people wanted to
hear what Steve had to say. Leda Cosmides was there. And there were
a lot of major figures like Ray Jackendoff and Daniel Dennett. On a
previous Tuesday night thing, Steve and Alan Prince had had a major
battle with Jay McClelland over computational models of language. It
was an astonishing intellectual event, and the graduate students were
talking about Steve's presentation many months later.

So Steve and I split our talk. Then Gould began his talk with, of all
things, a mildly offensive joke. He said something like, "I just got back
from a flight from Japan, and I'm exhausted—I got jet rag!" People
hissed.

Pinker likewise remembered the auditorium for the Tuesday night col-
loquium overflowing with people. "The crowd was far bigger than any previous audience at the series, and a partition had been taken down to double
the room size. They were all there to hear Gould.[8]

"But what Gould said," observed Pinker, "was surprisingly feeble. It
was clear that he hadn't prepared at all. He said something like, 'Well, lan-

guage can't be an adaptation for communication because it isn't always used for communication. For example, when I came here tonight from the airport, people asked me how I was, but they didn't really mean it.'"

The main thrust of Pinker and Bloom's argument was that it was obvious from the design of language that it had evolved: "If someone told you that John uses X as a paperweight, you would certainly be hard-pressed to guess what X is because all sorts of things make good paperweights. But if someone told you that John uses X to display television broadcasts, it would be a very good bet that X is a television set or is similar in structure to one, and that it was designed for that purpose. The reason is that it would be vanishingly unlikely for something that was not designed as a television set to display television programs; the engineering demands are simply too complex."

No matter how you look at it, they said, whether you consider organs that evolved to serve a specific purpose or something that started off as one thing, like a heat exchange, and then evolved to fulfill another purpose, a wing; there was no a priori reason that language could not have evolved stepwise like many other products of evolution.

It made as much sense to say that language evolved as a spandrel as it did to say that the eye could be some kind of architectural side effect of another kind of evolutionary change, said Pinker and Bloom. The reason you have all these very specific parts of the eye that perform particular jobs is because they evolved to do those jobs. Their jobs were their very reason for existence. The same is true for language. The rules of syntax and intonation and words matured over time into the system we have today because they were progressively refined by use and the forge of survival and reproduction—not because the brain got big and complicated for some other reason, and all of a sudden we discovered we could now manipulate symbols as well.

In addition to arguing for evolution from the design of language, Pinker and Bloom said there were many reasons why language could not be a Gouldian spandrel. Language was just too complicated. Spandrels are usually quite simple features. Even if a spandrel ends up being modified by evolutionary change and used in complicated ways, spandrels are typically "one-part or repetitive shapes or processes that correspond to simple physical or geometric laws, such as chins, hexagonal honeycombs, large heads on large bodies, and spiral markings."[9]

They reiterated that one of Gould's main problems with language evolution was that its supporters tended to rely on "just-so" stories, like the Rudyard Kipling tales that told how the leopard got its spots, to explain some critical developments. (Chomsky calls them "fairy tales.") In academia this is considered a term of abuse, and it essentially means you are making things up. The fear of being accused of fabrication was one reason that people stayed away from the issue of language evolution, said Pinker. But he and Bloom laid out many reasons why the evolution of language was a legitimate area of study. There are other compelling clues to the ways that language evolved; for example, our vocal tracts are shaped to produce speech, just as our hearing is specialized to register it.

Finally, they said, the argument against the evolution of language seemed to be based on nothing at all but the force of incredulity.

"Dan Dennett was there that night," said Pinker, "and more than once since then he has told me that he thought the debate was won by us. And yet as everyone was leaving, he was shocked to hear many people saying that Gould had clearly won." (Dennett was so incensed by this that he was inspired to write *Darwin's Dangerous Idea,* a bestselling book about the theory of natural selection.)

"After the talk," said Pinker, "we sent the paper to *Behavioral and Brain Sciences.* It had one round of reviews. We made the changes, and then it was accepted. We wrote it in 1989 and it got published in 1990, which for academia is fast."

Behavioral and Brain Sciences has an unusual format. When it publishes a paper, it includes comments from many other academics, to which the authors then get a chance to respond. Compared with just reading a single paper from a team of researchers (and then possibly reading a response to it two years later in another journal), it's a rich way to gauge the complexities of a subject and to understand what is at stake. The effect is of a dialogue.

The back-and-forth on the subject of the natural selection of language ran seventy pages: Pinker and Bloom's original paper was twenty pages, thirty-seven pages of comments from thirty-one different sources followed, and Pinker and Bloom responded in thirteen additional pages.

Many commentators were delighted by the paper. Jim Hurford, a linguist at the University of Edinburgh, who had been interested in the area of language evolution for some years, was thrilled. "I felt freed," he recalled, and aptly titled his reply "Liberation!" "Pinker and Bloom's target article is deeply satisfying," he wrote. "They correctly diagnose the consensus in linguistics and cognitive science, nurtured by the writings of Chomsky and Gould, that 'language may not be the product of natural selection.' Pinker and Bloom confront this stifling consensus head on."

The overwhelming impact of Pinker and Bloom's contribution stemmed not so much from the specific ideas about adaptation they proposed as from the stand they took against the idea that language evolution was an uninteresting or intractable subject. Working out the details of how language might have evolved remained a monumental task, but with their paper it was as if a door had been flung open. From that point on, more and more researchers felt that studying the origin and evolution of language was a legitimate academic inquiry. After a hundred years or so of uncomfortable silence, it had become intelligent, respectable, and interesting to wonder aloud how on earth we had come to be a species with words.

Influence isn't easy to define in academia. It may be obvious that a person or his ideas are powerful, but it can be hard to prove beyond simply pointing out that everyone seems to accept them. A more specific, if incomplete, measure of influence is counting how many times a scholar's papers are mentioned by other scholars in their own work. Yet another measure is the prestige of the journal in which the scholar publishes. (The influence of a journal is determined by how many times anyone cites papers it has published.) For instance, *Language,* the biggest journal in linguistics, has an Impact Factor (a measure of how often it is cited) of only 3. *Behavioral and Brain Sciences* has a score of 15.6, making it a powerhouse. In the case of Pinker and Bloom, although it's not possible to determine the relative contribution of all these factors, it's clear that together they had an impact. Before their paper, relatively few books and papers were published on the topic. Since then, many books and more than one thousand papers have been published on language evolution.

Why did the paper have such an impact? There's no guarantee that a clever, fascinating, and quite possibly correct academic article is going to be read. The products of science, like works of art, require intense focus and a lot of time to create, and then, typically, all but a few are ignored.

In part, the paper had the effect it did because of Pinker's stature. "Finally," Jim Hurford explained, "someone prominent, someone sort of in the Chomsky camp, someone generativist, was interested in language evolution."

Pinker agreed, "I think people liked it possibly because I was coming from so close to the politburo headquarters—being at MIT, where Chomsky was, and also just down the street from Gould at Harvard. Many people saw the paper as coming from someone who had no ideological ax to grind against Chomsky. I'm often seen as Chomskyan, even though I disagree with him on many things. But I'm close enough that that statement was all the more attention-getting."

Ironically, others were angered by the piece for the same reason—that Pinker was seen as an influential Chomskyan and yet he was disagreeing with Chomsky.

"Overall," said Pinker, "some people were grumpy. I think they were disgruntled because we contradicted the official line. And there was one exceptionally long and sarcastic letter written in response to our paper that was withdrawn from publication."

The language evolution paper also turned out to be a turning point in Pinker's development as an academic, for it got him started in evolutionary psychology. About a year after it was published, he started to think about writing a book for nonspecialists. In 1994 he published *The Language Instinct,* a prizewinning account of language as a biological instinct that hit the bestseller lists.

Bloom also did well. The *Behavioral and Brain Sciences* piece, only the second paper he had published, drew a lot of attention, and at the time he was on the job market—a fortunate coincidence for any graduate student.

Today Bloom is a professor at Yale and a successful author, as well as a co-editor of *Behavioral and Brain Sciences*.

After Pinker and Bloom, more and more people stopped asking, "Did language evolve?" and instead wondered, "How did language evolve?" Instead of being treated as an indivisible mystery, the problem of language evolution began to fracture into many good and answerable questions, like "What does gesture have to do with human language?" "How did categorical perception evolve?" "What's the relationship between music and language?"

In addition to its political impact, Pinker and Bloom's paper had the effect it did because they were writing about an idea whose time had come. Indeed, it was remarkable how many of the commentaries on their paper began with a remark like "Oh, how nice to see that Pinker and Bloom are now saying what I've been saying for twenty years. How nice that they agree with me."

It's true—while no one had previously enjoyed the attention that Pinker and Bloom got, a number of researchers had been toiling for years on the mystery of language and adaptation. By encouraging scholars to move beyond the Chomsky-Gould consensus, Pinker and Bloom not only inspired them to ask questions anew but created an opportunity for scholars to seek out earlier research on the topic and find out what had already been discovered beyond the borders of mainstream linguistic respectability.

4. Philip Lieberman

Because the light of evolution is not instantaneous or blinding, it is difficult to visualize the immensely slow and gradual change that is brought about by mutation and natural selection. When you consider a protozoan cell or an amphibian, on the one hand, and dolphins or, say, commuters, on the other, there is no intuitive way to make sense of the line that runs from one form of life to the next.

The popular cartoon of evolution, where an ape slowly unbends, straightens up, starts walking, and mutates into some form of modern-day human, is probably the easiest way to think about it. But as Stephen Jay Gould insisted, this caricature is misleading. Evolution does not follow the course of a single line. The tree of life bristles with stems, boughs, and branches. Most lines from one form to another are densely surrounded by branches leading to different species or to dead ends.

When it comes to the idea of language as an adaptation, the challenge of grasping evolution is further compounded by our inability to imagine ourselves without language. Language not only fills our lives, but we do our imagining, to a large extent, with language. Every now and then, we get a glimmer of what it might be like to exist without words. Sometimes there is a moment on waking when we are conscious but not self-conscious and our thoughts aren't shaped by language. We are looking up at the ceiling or across the room, and the ceiling or the objects in the room are just there, as we are there. We're awake but not much more. Is this what it's like to be pre-linguistic?

In addition to the natural obstacles to imagining how language, or anything, evolved, the way language was defined by generative linguistics made its evolution seem even more incomprehensible. Although Chomsky forswore explicit discussion of the language evolution question, many scholars thought the answer was implicit in his model of language. Indeed, Chomsky spoke often of innateness, and when you invoke innateness, it's hard not to make a few assumptions about genetics and evolution.

As a result, it seemed to many linguists and other cognitive scientists that the only way an innate universal grammar could exist, the only way humans could be born with a language organ, was if it was genetically endowed. The implication was that the language organ was specified in the genome, and generally it was assumed that there was a gene or genes specifically for language.

At the same time, Chomsky saw language as a perfect, formal system. So it appeared that a gene for this mathematical entity must have appeared out of nowhere with no precursors in other animals. This contributed to the widespread view that language evolution was impossible and language's very existence was miraculous.

Although Pinker and Bloom helped considerably to challenge that belief, some researchers had been resistant to this idea even earlier—Philip Lieberman, for example. Although Lieberman was once a student of Chomsky's, there is no interaction between them now. Both men are famously combative, and they have taken opposite positions on the subject of the evolution of language. In the 1980s and 1990s, while Chomsky expressed no interest in its study, Lieberman was examining skulls, listening to apes, and testing brains, all in search of clues to language's origins. Lieberman argues that not only should you study language evolution, but you can't even begin to understand language if you don't start with evolution. His research is grounded in the basic tenets of messy biology. When you look at the problem through his eyes, it becomes harder to see language evolution as either mystical or impossible. Instead, it looks merely insanely complicated.

Lieberman was born to a family of idealists and fix-it types. Both his parents had gone to the Soviet Union in the 1930s to save the world, but after a few of their Russian friends disappeared in purges, they left. Still, Lieberman grew up with books like *The Commissar of the Gold Express* lying about the house in Brooklyn. (He suspects his mother remained a sympathizer.) Lieberman's father, who learned his plumbing skills in the Soviet Union, ended up building highly classified plants for the atomic bomb project.

Lieberman himself completed a B.S. and an M.S. in electrical engineering at MIT in 1958, but after working on a few real-world projects for General Electric, he was bored with transistors and breadboards and decided to take a linguistics class. It was a low-key arrangement. There were only three other students, and the teacher handed out purple-ink ditto-machine copies of his notes on syntactic structures and transformations. The idea of transforming one syntactic structure into another by preordained steps had been around in linguistics, but in this class it was taken a step further. Transformations weren't just notational devices, said the teacher, but actual operations of the mind. It was the first linguistics class that Noam Chomsky taught.

Lieberman, who was twenty-two years old, found the class exciting, for he liked language and was intrigued by the idea of using it to understand the mind. Despite his enjoyment, however, his path soon diverged from Chomsky's. One day soon after his shift to linguistics, he wandered through the department—it was housed in a wooden building on campus where the first laser had been built—and was drawn by funny noises coming from a room off the hallway. He had heard the DAVO, one of the first speech synthesizers, and the engineer-turned-linguist became interested in how speech actually works.

He ended up writing his Ph.D. thesis (which eventually became the second book ever published by MIT Press) about how the physiology of breathing structures how we speak. Speakers make all sorts of muscular maneuvers in articulating words, and these are carefully controlled to make sure that the air pressure generated by their lungs stays at a steady level as they talk. Lieberman found that these maneuvers are keyed to the length of the sentence we *intend* to speak, showing that humans anticipate a long sentence before they utter a sound. The more he became engaged with these fundamental physical constraints on human language, the more he moved away from the abstract properties of language and toward all the things that Chomsky had dismissed as epiphenomena.

The problems of speech synthesis and voice recognition are far from solved today. When Lieberman began to wonder about speech, scientists were just beginning to get a glimmer of how complicated it was, and how enormously difficult it was to get a machine to either produce or understand speech. (One of the big differences between now and then is not that the problems have been solved but that researchers have come to appreciate the magnitude of the task.)

Once he started investigating the biophysics of speech, Lieberman only became more intrigued. The revelation that really shaped his future career came to him one night in the bath. After finishing his Ph.D., he got a job at the University of Connecticut, and one evening after work he lay in the tub, listening to WGBH. The presenter remarked that apes couldn't talk, and this struck him as worthy of investigation. Why not?

Lieberman often traveled to New York to teach at Haskins Laboratories and started spending time at Brooklyn's Prospect Park Zoo. When he took his tapes of hours and hours of ape vocalizations back to the lab to analyze, he found that apes do not make the full range of human sounds. This, he discovered, was because of the physiology of their tongues.

The human tongue extends from the larynx, deep in the throat, to just behind the teeth. At points along its length it can change its shape. It can be moved up, down, forward, and back; it can be bunched up or extended, widened or curled. Whenever the tongue changes shape, the whole vocal tract is altered, and each different configuration results in a different sound. In contrast, the tongues of other apes lie mostly in their mouths. As a consequence, they don't have the facility for generating as many specific sounds.

Lieberman also realized that even though there weren't as many sounds in the ape repertoire as in human speech, there were enough for the creatures to make a decent stab at talking. Chimpanzees can make *m, b, p, n, d, t,* and a number of vowel sounds. For a nonhuman, this is not bad. Few other animals can get close—if you could transplant a human brain into, say, a horse's head, it would not be able to speak human language, because its mouth and tongue could never make the sounds we do.

Where we differ from the chimpanzees is that they don't selectively articulate these sounds and manipulate their sequence, as we do when, for example, we say "pie," "my," "buy," "die," "tie," or "nigh." It is as if they have the same vocal instrument—or at least one that is reasonably similar—but they just don't use it in the same way.

If it was not the actual range of sounds produced by our respective vocal tracts that enabled us to speak but prevented apes from doing so, then, thought Lieberman, we must differ in our ability to control those sounds. This realization launched him on a quest to determine the connections between motor control and the higher levels of language. He quickly came to the conclusion that in order to truly understand language, you have to begin with biology, and—he is very fond of quoting Theodosius Dobzhansky, a famous evolutionary biologist who died in 1975—"nothing in biology makes sense except in light of evolution."

Lieberman's first book, *The Biology and Evolution of Language,* was published in 1984. In it he argued against the popular notion that there was a "linguistic saltation"—that is, no single dramatic event gave birth to human language. The Chomskyan idea of an ideal speaker and hearer confused the origins of language rather than illuminated them he said. Instead, he proposed: "Human syntactic ability, in [my] view, is a product of the Darwinian mechanism of preadaptation, the channeling of a facility that evolved for one function toward a different one."[1] He cited Darwin's discussion of the evolution of lungs from swim bladders: "The illustration of the swim bladder in fish is a good one, because it shows us clearly the highly important fact that an organ originally constructed for one purpose, namely flotation, may be converted into one for a wholly different purpose— namely respiration."[2]

Lieberman was not arguing, as he was careful to explain, that there was no uniquely human specialization for syntax. Rather, his point was that in the brain there was an overlap between the parts that control bodily movements and the parts that allow us to order thoughts and words in cognition and speech. This physical overlap had come about because of the way we had evolved, he said, first developing the ability to physically move our bodies in space and then, overlaid upon that, developing the ability to move words in abstract patterns.

All was peace and tranquility before the book, said Lieberman, but after its publication he and Chomsky fell out. For months, they argued back and forth, and then for the next eighteen years there was silence.

In 1990 Lieberman was invited by *Behavioral and Brain Sciences* to contribute one of the comments on Pinker and Bloom's paper. He wrote, "It is refreshing to see Pinker and Bloom adopting some of the major premises of my 1984 book: (a) that human linguistic ability evolved by means of Darwinian processes, (b) that the biological substrate for human linguistic ability is subject to the constraints of biology, in particular variation, and (c) that data from psycholinguistics, anthropology, neurophysiology, and so forth, are germane. However, Pinker and Bloom still carry much of the baggage of the MIT School of Linguistics, in particular that guiding principle 'Not invented here.' "

What he meant was that if research hadn't been done at MIT, then, as far as MIT was concerned, it didn't really exist. Clearly he was more annoyed than gratified.

Even though Lieberman, Pinker, and Bloom were all writing about language evolution, and even though they all agreed that any analysis of language needed to take biology seriously, there was at least one fundamental difference in their goals. Pinker and Bloom believed that Darwinian evolution and Chomsky's universal grammar were compatible, and sought to prove both Darwin and Chomsky right. Lieberman, on the other hand, believed the incongruity between slow evolutionary change and an innate language-specific organ was irresolvable. Pinker and Bloom's argument that universal grammar should and could take account of genetic variation was not acceptable, he said. In order to explore language evolution, you have to completely abandon the idea that humans are born with some kind of grammar device. It just wasn't possible for both Darwin and Chomsky to be right.

What Chomsky had wrong about language, according to Lieberman, fell into a larger category of misunderstanding biology. Throughout history, he argued, the most complicated piece of current technology was often used as an analogy for the human body or brain. For example, in the eighteenth and nineteenth centuries the brain was often thought of as a clock or a timepiece. It was imagined to be a telephone exchange in the early twentieth century. And from the 1950s onward, the brain was seen as a digital computer.[3]

These metaphors, Lieberman explained, often take on a life of their own. In the early nineteenth century, for example, physicians likened the body to a steam engine. When early steam engines became hot, they would explode, unless safety valves were used to release the intensely heated pressure inside. By analogy, doctors of the time bled patients who had a fever in the belief that releasing blood would lower the body's temperature.[4]

The human mind-brain implied by Chomsky's theory of language, Lieberman argued, was fundamentally based on the architecture and processes of a computer. In a computer, the central processing unit is a discrete device that generates output by algorithms. Random-access memory and hard drives are also modular mechanisms. The Chomskyan brain, similarly, has a localized language organ that generates syntax. Sound, structure, and meaning are constructed separately. And the language organ is separate from other parts of the brain, these parts also being separate from one another.

According to Lieberman, the analogy between the computer and the brain prevents a true understanding of language. Even though formulas can describe a set of sentences, they don't have much to do with how language is produced by the brain or how the brain and language evolved. "Syntax is not the touchstone of human language, and evolution is not logical," declared Lieberman. "Evolution doesn't give a damn about formal elegance."[5]

When Lieberman began working at Brown University in 1976, he turned his attention to the connection between higher levels of language and the motor system. He started with the basal ganglia. These neural structures, the striatum and the globus pallidus, lie beneath the cortex, the brain's outermost rind. The basal ganglia are responsible for learning patterns of motor activity—playing tennis, dancing, picking up a cup of tea. They also control the way different physical movements or mental operations are ordered, one dance step after another, and they are crucial in responding to a change in the direction of movement or thought.

Lieberman compared the basal ganglia of neurologically normal people with patients who had Parkinson's disease. In Parkinson's the brain progressively degenerates, and among the first and hardest-hit structures are the basal ganglia. The cortex is generally one of the last parts of the brain to be damaged, but when it is, the patient falls victim to dementia. People suffering from Parkinson's have tremors and rigidity and repeated patterns of movement. What intrigued Lieberman about these people was that they also had trouble comprehending and producing syntax. In addition to showing their physical symptoms, they tended to produce sentences that were particularly short, with only simple syntax.

Lieberman carried out a study of Parkinson's patients in which they were asked to say "one," "two," or "three" in order to identify which of three pictures best corresponded to a sentence they had heard. People who are neurologically normal generally make no errors when taking this test, but a number of the Parkinson's patients with damage to the basal ganglia struggled with sentences with slightly complicated syntax and with long, conjoined sentences of simple syntax.

In another study many Parkinson's patients were shown to have trouble if they first heard an active sentence ("The hawk ate the sparrow") and then were asked a related question in the passive voice ("Who was the sparrow eaten by?"). They also had difficulty when the original sentence was passive and the subsequent question was active. The patients experienced no problems in working out the meaning of sentences; it was just the syntax that tripped them up.

The fact that damage to a brain area that controlled motor skills also affected syntax was a smoking gun for a biological relationship between language and motor control. The basic idea, Lieberman argued, is that there is a dependent relationship "between the syntax of motor control and the syntax of language."

Interestingly, these findings overlapped with some of Steven Pinker's experimental results. Even though the two researchers began with opposite ideas about language and the mind-brain, they agreed on the subject of the basal ganglia and syntax. "Lieberman long ago predicted that the basal ganglia should have an important role in syntax," said Pinker. "And I found corroborative data that shows it." He continued:

A lot of my work on language uses a comparison between regular and irregular verbs as a way of tapping into the combinatorial, recursive part of language and the memory component of language. In particular, when we use "walked" as the past tense of "walk," you don't have to memorize that because you can just crank it out using the rule "add 'ed' to a verb." Whereas if you use "broke" as the past tense of "break," there you can't use a rule, because there is no rule. You have "break/ broke," but you have "take/took" and you have "fake/faked." So that relies on memory.

So comparing regular and irregular forms is a way of studying this recursive-combinatorial component in the simplest possible way— sticking an "ed" onto a verb is the smallest operation that anyone would be willing to call combinatorial or recursive grammar. The reason that the irregular is a nice comparison is that it doesn't involve a recursive-combinatorial component, but it means the same thing. It's just another way of expressing the past tense at the same length and same complexity.

We found that patients with Parkinson's disease have more trouble with regular than with irregular verbs, and they have more trouble with novel verbs. Like, when a new word enters the language, like, "to spam," everyone knows that the past tense is "spammed." I don't think you'd look that up in a dictionary or memorize it, but you can just deduce it from your world of recursive grammar. That's something that patients with Parkinson's disease have more trouble with than irregular forms, and that fits into Lieberman's theory that the basal ganglia are implicated in recursive syntax.

Lieberman has gone on to explore the basal ganglia in a completely different group of subjects. Starting in 1993, he began to compare the linguistic and motor performance of Parkinson's patients with that of individuals who were climbing Mount Everest. Both sets of people incur brain damage, specifically to the basal ganglia, though the basic cause is very different. Parkinson's is a progressive and fatal disease, whereas the basal ganglia damage suffered by climbers on Everest results from the lack of oxygen. In most cases it is temporary. Nevertheless, the climbers exhibit a lot of the same deficits experienced by Parkinson's patients.

Lieberman set up a monitoring unit at Everest's Base Camp, fifty-three hundred meters above sea level. His research team administered baseline cognitive tests to the climbers and took samples of their speech. As the climbers ascended the mountain and stopped at the next four camps, further tests and speech samples were obtained by radio link.

One of the abilities that Lieberman examined was how the climbers assembled the bits that make up distinctive sounds of speech. For example, when you pronounce *b,* you must coordinate at least two movements. At some point, you open your lips and release air while simultaneously vibrating the vocal cords deep in your throat. Timing the onset of voicing in speech sounds is yet another complicated motor skill at which every normal speaker is expert, though few are consciously aware of it. It is also another kind of movement sequence that gets affected in Parkinson's disease.

For example, the only difference between a *b* and a *p* is that you vibrate your vocal cords much sooner for the former than for the latter. With a *b,* voicing occurs within twenty-five milliseconds of opening your lips; with a *p,* your vocal cords start vibrating more than twenty-five milliseconds after you open your lips. Because Parkinson's patients experience a breakdown in the onset timing of voicing in speech sounds, some of their *b*'s sound like *p*'s, and vice versa. (The same applies to *d* and *t* and to *g* and *k*.) This deficit occurs alongside an increase in syntactic errors and a delay in the comprehension of simple sentences.

Lieberman showed that the higher the climbers went up the mountain, the more trouble they had with the timing of their voicing and the more their comprehension of syntax degraded. The farther up they went, the less oxygen they breathed, and just like Parkinson's patients, they became less adept at distinctly pronouncing sounds like *b* and *p,* and they took longer to understand test sentences.

It's clear from this evidence, according to Lieberman, that the basal ganglia are crucial in regulating speech and language, making the motor system one of the starting points for our ability not only to coordinate the larynx and lips in talking but to use abstract syntax to create meaningful and complicated expressions.

One of the important functions of the basal ganglia is their ability to interrupt certain motor or thought sequences and switch to a different motor or thought sequence. Climbers on Everest become increasingly inflexible in their thinking as they ascend the mountain—stories about bad decision making in adverse circumstances abound. Accordingly, Lieberman's climbers showed basic trouble with their thinking.

One mountaineer monitored by Lieberman scored well at base camp but demonstrated extreme anomalies in his speech and a dramatic decline in thinking as he ascended. The researchers told him that he wasn't functioning normally and advised him to descend, but he refused, insisting he was fine. When the weather took a turn for the worse and his companions descended, he persevered in going forward. A few days later he fell to his death.

It was later discovered that at the time of his death, a harness the climber needed to secure himself to fixed ropes was not properly attached. There was nothing wrong with the harness itself; the problem was in how it had been used. In order to secure the harness, a correct sequence of steps had to be carried out. It appears that the lack of oxygen supply to the basal ganglia affected the climber's ability to follow the basic sequence of clipping and unclipping.

Basal ganglia motor control is something we have in common with many, many animals. Millions of years ago, an animal that had basal ganglia and a motor system existed, and this creature is the ancestor of many different species alive today, including us. When we deploy syntax, Lieberman argued, we are using the neural bases for a system that evolved a long time ago for reasons other than stringing words together.

Chimpanzees, obviously, have basal ganglia. Birds have basal ganglia. So do rats. When rats carry out genetically preprogrammed sequences of grooming steps, they are using the basal ganglia. If their basal ganglia are damaged, then their separate grooming moves are left intact, but their ability to execute a sequence of them is disrupted. (Lieberman calls their grooming pattern UGG, universal grooming grammar.)

The fact that a number of different animals use the basal ganglia for sequencing, whether it involves grooming or words, said Lieberman, suggests that there is no innately human specialization for simple syntax. Instead of being a contained and recent innovation in the human lineage, the foundation of syntactic ability is an adaptation of our motor system, a primitive part of our anatomy.

Lieberman's contrarian (at least prior to 1990) take on language and its history offers an entirely different way of thinking about language evolution. When he started engaging with the subject of language, he wrote of it as not so much a new thing that humans *have* as a new thing we *do,* and we do it with a collection of neural parts that has long been available to us. Moreover, when you think about language this way, it is not really a "thing" at all but a suite of abilities and predispositions, some recently evolved and some primitive. The many parts of the brain and body that make up the language suite allow us to program into our own heads how our parents speak. When Lieberman calls language part-primitive and part-derived, he echoes Charles Darwin, who wrote in *The Descent of Man* that language was half art and half instinct.

The nineteenth-century German philosopher Arthur Schopenhauer said: "All truth passes through three stages. First, it is ridiculed. Second, it is violently opposed. Third, it is accepted as being self-evident."

The study of language evolution from the nineteenth century onward has rather neatly followed the same course as Schopenhauer's aphorism. Linguists once considered pursuing the topic an absurd endeavor. Then it was banned. After that, the official ban developed fairly seamlessly into a virtual ban. Now, where most researchers once glibly proclaimed that you can't study it, many say you can, including the scholar best known for saying you can't (or at least, you shouldn't bother).

We are at a strange now-you-see-it-now-you-don't moment in the history of language and mind where it seems that everyone is taking possession of the same new attitude. It's remarkable, now that the rhetoric about language evolution has shifted, how quickly what was once heretical has become received wisdom. Within a few years, students in Linguistics 101 will probably assume that asking about language evolution was always this easy and obvious.

In a relatively short time, academics like Savage-Rumbaugh, Lieberman, and Pinker, in their different ways, have had enough influence to make the subject no longer controversial or taboo but a legitimate line of inquiry—an endeavor about which reasonable people could disagree.[6] When questioned about the investigation of language evolution at the 2005 Morris Symposium on the Evolution of Language, Chomsky himself shrugged his shoulders and said, "I wouldn't have guessed it could go so far."

Of course, there are still profound disagreements among the researchers. Even though Chomsky published a paper that discussed language evolution in 2002, he remains immensely discouraging about the subject. In addition, he argues that it is possible to engage with language evolution for purely logical reasons that are internal to linguistic analysis. Pinker and Lieberman, on the other hand, build their respective cases with findings from genetics, psycholinguistic studies, and experiments that compare the cognition and communication of various animals and humans. However, they disagree completely about the nature of syntax. In 2003 Sue Savage-Rumbaugh announced that Kanzi had uttered his first spoken word, but of Chomsky, Pinker, and Lieberman, only Lieberman considers her work to have crucial insights for language evolution.

Nevertheless, the findings of all these scientists are important touchstones, and thanks to their disagreements, engagement, and even disengagement, the field has widened considerably. A great multidisciplinary conversation is now taking place. American biologists, Italian physicists, Australian neuroscientists, British anthropologists, and a variety of linguists and computer scientists are but a few of the academics investigating the origins of language. Researchers like Marc Hauser and Tecumseh Fitch, who co-wrote Chomsky's 2002 paper on language and evolution, are proposing an entirely new field of study—evolutionary linguistics.

The consensus among these researchers resonates with one of Savage-Rumbaugh's and Lieberman's main points: that what evolved was not a single thing. Language is not a monolith.

Part 4 of this book will delve more deeply into the disagreements provoked by the new questions, but first we will survey the diverse array of experiments that purport to explain what the components of language are and what processes were responsible for drawing them together. Lieberman showed that the basal ganglia are implicated in the evolutionary trajectory that led to language, but what other parts of the brain are involved? What do they contribute? What about thought? What about gesture and speech and words? How ancient are they? And what relationship do genes have with language?

Michael Arbib, a neuroscientist who investigates mirror neurons, thought now to be important for language, prophetically claimed that as the field develops, "we are going to be dazzled and puzzled and infuriated by the number of ways that language is defined."

II. IF YOU HAVE HUMAN LANGUAGE . . .

Rats execute one movement after another in a logical sequence when they groom themselves; it doesn't mean they can tango or, for that matter, rap. Human linguistic ability, taken as a whole, is still completely unlike anything else in the biosphere—which is why most of the key skirmishes in language evolution revolve around the issue of uniqueness. Does language make us unique? Is language unique to humans? If rats possess the same sort of systems that we use for language, does it make us less unique? Does it make us more like rats?

These are reasonable questions. Obvious and enormous distinctions exist between human life and other animal life on this planet. We communicate with animals, but we don't converse with them, and they don't talk to each other. They don't read magazines, they don't write books, and they don't compose poetry. Language appears to be uniquely unique. Naturally, we want to explain why this is so.

But asking what makes humans unique is almost always qualitatively different from asking what makes the antelope unique, or the sloth, or the dung beetle. These questions don't have to be different, but they have historically been so. The former is never purely scientific, but is inevitably shaded by our self-regard and is always, to some degree, existential. We think that working out what distinguishes us from them will entail working out what makes us *us*.

Throughout history, many "uniquely human" attributes have been proposed. We make tools. We are creative. We have culture. We play. Inevitably, these characteristics get disputed. Chimpanzees use tools, crows design and build them, half a dozen species have culture, practically everything plays. Nevertheless, we keep trying to draw an absolute line in the sand.

Anthropocentrism crops up at every level of research. Science as a body reeled when it was announced in 2001 that the human genome consists of maybe 30,000 to 40,000 genes, only 10,000 to 20,000 more than a roundworm's. It was always assumed that if organisms as simple as worms had about 19,500 genes, we must have many times more—at least 120,000 of them. Since the 2001 announcement, the estimate has dropped even further, and the same group that put the human genome at 30,000 to 40,000 genes revised the number downward in 2004 to 20,000 to 25,000.[1]

For the same reasons that the size of the human genome was wildly overestimated for years, animal cognition has been prone to underestimation. Indeed, it is only very recently that the word "cognition" has even been used to describe animal thinking, planning, memory, and knowledge. Frans de Waal draws a parallel between the study of animal cognition and the topic of animal emotion: "Emotion is basically taboo still. For example, there's a scientist I know who was one of the pioneers of empathy studies in children, and she told me that twenty-five years ago, if she presented anything on empathy in children, she would be categorized with groups like people who study telepathy. I think in animal studies we are now at that stage where she was twenty-five years ago." He adds, "If I mention emotions in animals, a lot of scientists become very squeamish.[2] So that's still a taboo topic, and it has the same flavor as the cognition topic. There used to be objections to cognition studies, because people would say, 'You cannot look into the head of an animal.' You also cannot look into the head of a human."

De Waal's book *The Ape and the Sushi Master* (2001) is an insightful account of the history of anthropomorphism and anthropocentrism in animal research, which he believes is still relatively widespread in science. In 2002 he said:

> Well, there's still a large category of scientists who would strongly object to comparisons between humans and other animals. They don't just want to keep humans and animals separate; they also want to keep them separate in the language that they use. It's almost as if the world is divided into two kinds of people. There are certain people who ob-

ject to the comparison with animals and feel insulted by it, and there's another group of people who think it's fine and, "What's the big deal?" You see that in the sciences, but you also see it in philosophy and you see it in authors who write. You see it everywhere.

Like studies of cognition and emotion, language research is one of the last areas in which ideas about what makes us biologically unique and what makes us personally special still get muddled. It's not unusual to run across a statement like the following: "One of man's greatest achievements is language"—as if all the species took an exam and humanity was the only one to pass. (No one talks about the grand successes of the fruit bat or the accomplishments of the pygmy owl.) It's not hard to infer from statements like this that we had some kind of conscious control over the evolution of words and rules. But there is no agency in evolution; it is inadvertent. We survived, modified, and multiplied, just like any animal alive today, and out of the wildly dodgem course we took, language arose.

The intimacy of our relationship with chimpanzees has been apparent from the fossil record for some time, but in recent years it's been underscored by genetic evidence. It's now a much-quoted truism that our DNA sequence is, on average, 98 percent the same as that of chimpanzees (and only marginally less than that with other primates). Defining what it is to be human has, for many researchers, meant looking only at the 2 percent difference and assuming that in that small gap lies the key to all the cities of the world, all the farms, the oil rigs, the fast-food franchises, all of culture, and language.

This plays out in many different ways. For example, scientists assumed for a long time that the parts of the brain that have to do with language must be wholly new, recently evolved additions that we do not share with nonhumans. The underlying idea here is that we attained a superior enough level of intelligence or complexity to acquire language, a sort of evolutionary benediction.

Because researchers who take this approach see little in common between human and nonhuman communication, investigating the linguistic overlap between us and other animals has traditionally been dismissed as wrongheaded or irrelevant. In this view, the implication also lurks that human language can be acquired as long as you have a sufficiently powerful brain, as opposed to a specifically *human* brain shaped by specifically *human* constraints. But if you could somehow make the brain of, say, a crocodile smarter in some general sense, it would not automatically burst forth with human language.[3]

So if you happen to have human language, it means that you are . . . human. You are terrestrial, you are a mammal, you are bipedal, you are a social primate. And if you are human, your ancestors traveled a particular evolutionary path, and many animals alive today have ancestors that walked much of the same path.[4] In this zoomed-out view of humans as animals, discounting the communicative abilities of other animals, argue scholars like Lieberman, is at best hermetic; at worst, it's unscientific. Which of these traits are necessary for human language and which are incidental can be determined only if they are all considered in the first place.

Although we have many ancestors in common with other species, we have more recently evolved alone. So how do you explain both continuity and uniqueness? The scholars who are trying to break down the monolith of language and work out where the seams lie seek the essence of humanity and of language in both the 2 percent and the 98 percent. Yes, language as we know it today is uniquely human, but, says Lieberman, "human linguistic ability can be traced to the motor response of mollusks."

Evolutionary biology takes the view that all animal kind, despite its gorgeous multiplicity, is merely a set of variations on a theme: life. You are cousin to the world's roaches, puppies, and Tasmanian devils because you share an ancestor, that once-upon-a-time cell that winked, split, and got the whole thing rolling.

The continuity of animal life is not just an intellectual orientation or frame of analysis; it's a visceral, involuntary experience. In Leipzig, Germany, in 2004, a group of scientists got together to discuss gesture and language evolution. In one presentation, baboon communication was discussed. When you are a male baboon, communication can be very high stakes: it's not uncommon for one of these animals to reach forward and rip the testicles off its interlocutor.

When the presenter described this unpleasant form of exchange, a flush spread through the conference audience. Everyone present was a scientist of some sort, most worked with animals, yet the joking and laughter and the slight sense of hysteria that followed was semi-involuntary. The remark would not have had quite the same intensity if the animal in question was a bug and the body part an antenna. Human life is continuous with all life, and baboons are pretty close to us on the continuum.

One way biologists measure traits is by asking whether they are shared with other animals. If a trait is shared by two or more species, the next question is whether the relationship is one of homology or analogy. If the traits shared by two or more species are homologs, then they have come to those species from a common ancestor, perhaps an ancient one, which also had the same trait. Human arms and bat wings are homologous. And you can cut the distinction finer: if a trait is homologous between two species, it's useful to ask if it is effectively the same or whether there is only a partial similarity.

If a trait is an analog, then even though it is shared by different species, the trait evolved separately in those species. Biologists say that these species have converged upon a similar solution to a problem. Bat wings and butterfly wings are analogs. A trait that is analogous between species may or may not be especially old in either one or both of them.

Analogy is a useful concept because it demonstrates that it's possible for the same trait to evolve even in very different species. For example, both humans and songbirds are exceptional at learning vocally—humans with language and song, and songbirds with the songs they grow up hearing. Very few species can match their skills in this respect. One useful observation that results from comparing the two is that you don't have to be human to develop a particular expertise—even if that expertise is part of what makes us human.

It's important to be cautious when labeling a trait homologous or analogous. For example, humans think of themselves as tool users par excellence. But other great apes, such as chimpanzees, gorillas, and orangutans, are also skillful with tools. The mother of all the great apes, the ancestor of orangutans, gorillas, humans, chimpanzees, and bonobos, lived about fourteen million years ago. Over time, the descendants of this grand-primate diverged into two species: one that evolved into the modern orangutan, and one that was the common ancestor of gorillas, humans, chimpanzees, and bonobos. This latter creature lived for around seven million years before its line also split into two paths, one of which led to the modern gorilla and one that led to humans, chimpanzees, and bonobos. The common ancestor of humans, chimpanzees, and bonobos lived about six million years ago, after which the line that led to humans split away. The human lineage threw up at least twenty different types of hominids, only one of which survives today, *Homo sapiens sapiens*. The other line split again a million or so years later and led to the two chimpanzee species that exist today, *Pan troglodytes,* the common chimpanzee, and *Pan paniscus,* the bonobo, found only in Zaire.

The branching of the great ape family tree means that chimpanzees and bonobos are our closest relatives on the planet. Even though to the human eye a bandy-legged chimp may look more like a gorilla than like a human, the chimp is actually more like us than it is like the gorilla. As noted by the evolutionary biologist Jared Diamond, humans are really just the third species of chimpanzee. Perhaps to a gorilla, we all look alike.

Based on how closely related we are to chimpanzees, you might assume that they would be the best tool users of the other great apes, and although there is good tool use data for them (as well as gorillas and orangutans), that's not their reputation. There is a saying among primate keepers, Heidi Lyn explained, that if you give a give a screwdriver to a chimp, it will throw it at someone. If you give a screwdriver to a gorilla, it will scratch itself. But if you give a screwdriver to an orangutan, it will let itself out of its cage.[5]

If traits are not shared, then it usually means they are species-specific. Language, as in the whole suite, is clearly species-specific—it is not shared and has no homologs or analogs on the planet. But what happens when you disassemble the monolith? To what degree is this new function made up of old parts? And how old is old? Do some parts share an evolutionary trajectory? And how many of them are shared? Are they completely interdependent or somewhat independent of each other? How many, if any, are uniquely specific to humans?

5. *You have something to talk about*

Twenty-two New Caledonian crows are housed in the aviary at Oxford University's zoology department. The birds inhabit two small rooms, each subdivided by a wall. A hatch allows the crows to pass from a closed room to another where a wire mesh lets in fresh air and sunlight. The birds' feathers are thick and glossy, and their claws are three inches from the tip to the back. Long sticks are strung up as perches from floor to ceiling in the aviary, and the birds hop constantly from perch to perch, displacing one another like three-dimensional dominoes. If their bodies are at rest, then their heads are in motion, swiveling up, down, and around, either to survey the surroundings or to preen the feathers of the crow next to them. On the island of New Caledonia in the southwest Pacific the birds are plentiful, inhabiting terrain from rain forest to mountain. They are long-lived and have complicated social lives and a wide range of different vocalizations.

The crows are also shy, so when the graduate student Ben Kenward enters one of the rooms of the aviary, they disappear through the wall hatch. On a table in the birds' room, Kenward places a test tube that is held sideways in a clamp. Inside the tube are tiny chunks of meat; next to it Kenward lays out small kebab sticks. When the crows return a few minutes after he has left, the most confident one sails straight from perch to table. It picks up one of the skewers in its beak and without hesitation inserts it into the tube, poking and pulling, until it manages to roll out a piece of meat. Meat in mouth, the bird returns to a perch to enjoy its food. The scene replays itself when another bird steps up, this one plying the stick until all the meat is gone. The crows love their sticks. Even after the food is consumed, they hang on to them, fly them around, drop them, pick them up again, and chew strips off them.

The aviary's longest-term tenant is Betty. In 2001 Betty was filmed by the lab researcher Alex Weir, then a Ph.D. student, who wanted to see if she or her aviary mate at the time, Abel, would intentionally choose a hooked tool over a straight one to get food in a test in which only the hook would work. He presented the birds with a glass cylinder inside of which was a tiny toy bucket with handle erect. Inside the bucket was meat. The only way for the birds to get the meat was to remove the bucket from the cylinder, and the only way to get the bucket was to use the hooked tool. In one of the very first trials, Abel accidentally knocked the hook away, leaving only the straight tool. Betty quickly hopped up after Abel and in a completely businesslike fashion took the remaining straight piece of wire—a material she'd never seen before—found a suitable place to wedge it, bent it into a fine-looking hook, and then used it to retrieve the bucket and then the meat.

One fundamental idea shared by many researchers is that in order to evolve language you first have to have something to say—as opposed to, for example, going about your life, developing language out of the blue, and then finding you have a lot to talk about. The search for the origins of language thus includes a search to uncover what ultimately was so worth relaying that our ancestors began to ratchet up their communication skills in order to do so. In trying to work this problem out, it helps to know what kind of thought goes on in the heads of nonlinguistic creatures. For a long time, we have assumed that not much does.

This conviction comes in part from the human tendency to believe that all of our complex ideas and ways of carving up the world are a result of the fact that we have language. Indeed, it can be hard to imagine otherwise. Everyone reading this book has probably experienced the odd sensation that he or she was momentarily without words, a state that mostly feels like a vacant one. Likewise, few people would claim they could remember what thinking was like before they learned language. The sense that not a lot was going on at that very young age is probably fairly common. Accordingly, we presume that our pre-linguistic ancestors had a pretty simple mental life.

But how do we account for Betty? She is a completely languageless creature, and yet she's no stranger to brilliant and rapid thinking. Betty not only saw that a hook was necessary to lift the bucket in order to get to the meat, she didn't even try the straight wire first to see if it would work. She simply went about creating the tool she needed to reach her goal. Professor Alex Kacelnik, head of the Behavioral Ecology Research Group, likens her act of inspiration to the way a chess grandmaster makes a decision after viewing a given situation on a board and consciously examining only two or three moves out of the range of possible ones. Something is going on in the back of the player's mind that leads him to reject all but a few choices, said Kacelnik. It's a creative process that he may experience simply as an aesthetic judgment. In contrast, a computer, Kacelnik explained, must trawl through all permutations of possible chess moves. In this respect Betty is like the human, in that she must have engaged in some kind of planning that involved unconsciously discarding useless strategies and focusing on a successful one in order to spontaneously produce the hook.

Until Betty's invention was videotaped, no other animal aside from humans had been shown to build its own tools, substantially altering a basic material to an appropriate design. In fact, it wasn't so long ago that we believed tool use was uniquely human. This changed when primatologists showed it was a perfectly normal activity for nonhuman primates. Apes fish for termites with sticks, crack nuts with stone and anvil, and process plants for food by whacking the tops of palm trees with large fronds. Scientists announced at the end of 2005 that gorillas in the Republic of the Congo were observed using sticks to test the depth of water before they stepped into it. (Earlier it had been assumed that gorillas were the only great apes that did not use tools at all.) In early 2007, an Iowa State University team announced that chimpanzees in southeast Senegal were observed sharpening sticks and using them as spears to hunt bushbabies. Also recently announced was the finding that dolphins off Australia's west coast use sponges on their snouts when they probe for food on the seafloor. These animals show that tool use is not only *not* restricted to linguistic creatures and their close relations, it doesn't even require two arms and two legs.[1]

In 2003 the New Zealand researcher Gavin Hunt announced that New Caledonian crows in the wild create different types of tools from the island's pandanus leaves. Like humans, said Hunt, the crows sculpt raw material into distinct tool designs with a highly regular shape. In other words, they don't just pick up a stick that's most likely to work for a job at hand and simply use that. Their tools are purpose-built. Additionally, the crows pass on their techniques for distinct designs to other crows.

As the attitude toward animal cognition has changed since 1990, more and more research in the area has shown that, as with Betty, there is some pretty complicated mental processing going on in the heads of animals that do not have human language. For students of the evolution of language, these investigations help to differentiate between the mental platform that our species may have had before we had language and the kinds of thinking we do now that are shaped by the fact we have language—the cogitation that is more directly part of the language suite.

Sometimes ideas are turned over in science when someone stumbles across one crucial counterexample—like a gorilla using a tool. In other instances the dismantling of assumptions is a conscious effort. Such is the case with the bird brain's recent upgrade. For a long time, neuroscientists conceived of bird brains as basic instinct machines. But in 2005 an international group of scientists calling themselves the Avian Brain Nomenclature Consortium announced a new set of terms to describe the bird brain. Their proposal was more than just a substitution of vocabulary; it represented a newly sophisticated understanding of a brain that had until then been sorely underrated.[2]

Human brains have a neocortex, a relatively recently evolved sheet of neurons that encases the brain. It is in the interactions between this area and the older segments of the brain that a lot of our processing is done. Birds have no neuron sheet, and this led researchers to assume that few comparisons could be made between avian and human brains. However, scientists eventually realized that birds possess a neural module that is functionally equivalent to the human neocortex.[3]

For a scientist like Irene Pepperberg, who has been engaged in avian behavioral experiments and observations for decades, the consortium's findings are a vindication: birds are much smarter than they are given credit for.

Alex, a thirty-year-old African gray parrot, is the most famous resident of Pepperberg's lab at Brandeis University. He has been filmed by camera crews from all over the world and appeared in stories in major newspapers. Pepperberg has written a book about him, and he was recently featured on *Scientific American Frontiers*. ("He loved Alan Alda," said Pepperberg.) For decades Pepperberg has been teaching elements of English to Alex, who has the language capabilities of a two-year-old and the cognitive capacities of a six-year-old. He can explain his needs and wants by using language.

Alex is about twelve inches from beak to tail, and he weighs only one pound. His companions are ten-year-old Griffin and seven-year-old Arthur. He has a clean white face, soft gray feathers in differing shades that are delicately scalloped around his face, and an intensely red tail. The lab where he lives is fairly small, about 150 square feet, with cinder-block walls painted white to cheer the birds up and newspaper spread all over the floor. The birds sit on perches in front of their cages. Otherwise there is room only for Pepperberg's small desk and a set of shelves that store the birds' experimental materials: plastic letters, colored stacking cups, and wooden blocks. For twelve hours the birds sleep in complete darkness, as they would in equatorial Africa. The rest of the day they always have at least one human companion to watch over them and work with them.

All the birds eat a specially formulated pelleted diet, supplemented with shredded wheat, colored pasta, vegetables, and fruits, and when Pepperberg offered Alex a piece of a muffin, he accepted it with a "Guuur-rrrrrreat!" and then "Yummy." He calls it "banari," a combination of "banana" and "cherry"; it is his word for "apple," explained Pepperberg. Alex's voice is distant and tinny, like a recording from an old-style Victrola. Because Alex has lived all over the country with Pepperberg, his "carrot" sounds Midwestern, while his "shower" is Bostonian.

Alex can identify by word fifty different objects, seven colors, and five shapes. He comprehends numbers under ten (though he doesn't count sequentially, he may use counting for quantities above it), and he can make distinctions between things that are the same and things that are different. Once he has learned new words, Pepperberg tests him on them. She fills a tray with blocks—maybe four green and two blue—and asks him, "How many blue?" Alternately, she'll ask, "What color two?"

One of Alex's most recent accomplishments was learning to transfer his concept of "none" from the same-different study to numbers. "Folks have studied the concept of zero in chimpanzees, but never in birds," Pepperberg explained. "None" is considered a particularly sophisticated concept for humans. "What I'm finding," she said, "is that Alex can use 'none,' without training, to refer to an absence of quantity in some situations. So, if I give him a tray of two blue, four green, and six yellow blocks and ask 'What color five block?' he'll say, 'None.'" What's most surprising about the fact that Alex understands what "none" means is that he was trained to use "none" when asked what was the same between a set of objects when in fact nothing was the same. (He was also trained to use "none" when asked what was different between a set of objects when nothing was actually different.) "Alex spontaneously used 'none' to denote the absence of difference in size between a pair of objects, and then also spontaneously transferred it to the 'What color x?' task. I had a tray of blocks and was asking, 'What color four?' and he kept saying 'Five.' I was pretty frustrated, and without thinking I finally said, 'Okay, what color five?' to which he replied, 'None.'"

Alex's understanding of "none" is more like a child's than an adult's: "If I show him that nothing is hidden under a cup and ask him, 'How many nuts?' he is like some autistic children or like children around three years of age. He simply refuses to answer. For him there is nothing there to comment on."

In a demonstration Pepperberg sat Alex near her desk and showed him various trays of blocks. She got him to identify the colors and amounts. Then she put the testing aside and tried to teach Alex the color white, a new category for him. She held a square piece of paper up to him and asked, "What shape?" "Corners," said Alex. "Yes, it has corners," she said. "What color?" As he did with every other object she proffered, Alex beaked the paper with interest. He stalled for a few minutes, then said, "None." Pepperberg burst out laughing. "Okay, fair enough," she said. "In your world, this has no color."

I asked Pepperberg if she thought the ability we share with Alex to use these categories of number, color, and shape in making sense of the real world results from convergent or direct evolution. Is it possible this ability goes back as far as the remotely distant common ancestor shared by humans and African gray parrots? For now, it's not clear, she said, but the amount of neurological and neuroanatomical evidence is growing. The abilities might be homologous, but at this stage the possibility is speculation only. "I think we are at an incredibly interesting point where we're beginning to learn more about both human and animal systems," she said. "The amount of knowledge we're going to gain in the next years is going to be exponentially greater than what we've learned over the past few years."

Alex's talents demonstrate that not only is the ability to understand and act on general conceptual categories like color and shape and number not human-specific, it's not specific to apes, or even to mammals. Alex can use those categories in the comprehension of complicated labels, and in the larger meaning created by stringing some of these labels together, like "What color five?" We may have words for these concepts, but it's clear that you don't have to have language to understand them and to be able to act on that understanding.

Other sophisticated forms of cognition include awareness of oneself and the ability to generalize. Gordon G. Gallup started exploring self-awareness in animals in the late 1960s, when he began to look at the way animals make use of reflection. Different animals interact in different ways with mirrors. Some ignore them entirely. Others use mirrors to locate things in space; parrots, for example, can find hidden objects that are visible only by their reflection. Other animals, like monkeys, engage with their reflections as if the reflection were another individual entirely. Gallup was the first to show that chimpanzees recognize that the image they are looking at in a mirror is themselves, an ability that was previously thought to be human-specific.[4] When Gallup announced his findings, many researchers were shocked and unsuccessfully tried to disprove them.

In 2000 Diana Reiss, at Osborn Laboratories of Marine Sciences at the New York Aquarium in Coney Island, and Lori Marino, a senior lecturer in the Neuroscience and Behavioral Biology Program at Emory University, applied the test to dolphins. Like all other whales, dolphins have traveled a radically different evolutionary trajectory from ours. Their closest land relatives are the ungulates, like the hippo. The researchers marked the dolphins with a nontoxic black marker on parts of their bodies that couldn't be seen without the use of a mirror, and then watched and recorded their behavior at a mirror attached to the outside glass wall of their pool. Once the dolphins had been marked, they swam to the reflective surface and used it to examine the ink marks, showing clear awareness of themselves. In fact, Reiss pointed out, the test didn't just expose the capacity for self-awareness: it also demonstrated that the dolphins were motivated to view themselves.

In 2006 it was announced that elephants are able to recognize themselves in mirrors. This work was also conducted by Diana Reiss, with Frans de Waal and Joshua Plotnik. In a similar fashion to the dolphin experiments, the researchers marked three Asian elephants and gave them access to mirrors. They noted that, compared to dolphins, elephants have the advantage of being able to touch most of their body with their trunks. Accordingly, after being marked the elephants spent a significant amount of time in front of the test mirror, repeatedly touching the experimental marks (but, crucially, not touching invisible marks that had also been made).

Reiss's work with dolphins has also provided evidence for the ability of nonlinguistic animals to generalize. Dolphins instinctively eat only live fish, so in captivity they must be taught to consume prey that is already dead. Reiss had to cut each fish she fed them into three parts. A dolphin would happily eat the head and the middle, but it would eat the tail only if the fins were cut off. If the dolphin misbehaved during feedings, Reiss gave it a time-out. This involved getting up from where she knelt at the side of the pool, walking back about twenty feet, and looking at the dolphin but not interacting with it in any way for a minute or so. "It let her know something was not right," explained Reiss. One day Reiss accidentally let an untrimmed tail slip into the dolphin's food. The dolphin responded by swimming to the opposite side of the pool and then rising out of the water in a vertical position, just looking at Reiss for a minute or so. *This feels a lot like a time-out!* thought Reiss.

She decided to test the dolphin, and a few days later she let an uncut fish tail slip through on purpose. The dolphin did the same thing, giving her another time-out. Reiss repeated the experiment three additional times, each with the same result. Dolphins are natural imitators, said Reiss, and imitation is an important part of the ability to learn. They are what Reiss calls "contingency testers," forever probing and exploring objects, and extremely adept at recognizing and generating patterns. The intentions behind their actions can be as obvious as our own.

Chimpanzees are also known to be good at generalizing and applying the patterns of one task to another. It is this ability that makes them such exceptional subjects in cognitive experiments. Monkeys, in contrast, can't generalize. They may be close relations, but if they learn how to use a joystick in one experiment, they have to relearn how to use it for the next.

The ability to grasp the concept of number, like most other mental talents, was believed to belong only to speaking humans, until researchers began to explore it in babies and other animals. Since these investigations began, the evidence for a shared, fundamental comprehension of numbers has mounted. Babies are able to identify numbers below four exactly, and they can represent large numbers approximately. In 1992, in one of the first experiments of this kind, the researcher Karen Wynn showed infants a Mickey Mouse doll and then hid it behind a screen. Wynn then showed the children another doll and placed it behind the screen as well. When the screen was removed, children were startled if they saw only one doll, and they looked longer at the object. Further experiments demonstrated that children were able to understand the addition and subtraction of up to three objects.[5]

As research on the natural abilities of infants has accumulated, so it has for animals. In 1999 two researchers at Columbia University announced that they'd taught two rhesus monkeys to count to four using images of shapes on a computer screen. The monkeys were also able to understand the difference between smaller and larger numbers with greater sets of images.[6]

Marc Hauser, who is head of the Cognitive Evolution Laboratory at Harvard, and his colleagues have shown that monkeys can, like children, grasp small numbers precisely and approximate large numbers. They can also perform the same kind of addition that babies can. Hauser replicated Wynn's experiment, but instead of human babies he used rhesus monkeys as subjects. Like the children, the monkeys were startled when the numbers didn't correctly add up. In later experiments the researchers further investigated the ability of the monkeys to understand addition and subtraction of amounts up to three. These findings also held true for domesticated dogs.

In 2006 French scientists announced that children and adults from the Munduruku, an isolated group of indigenous Amazonians, had demonstrated that they understood and were able to use concepts from geometry even though their language has no words for those concepts. When investigators showed them drawings of parallel lines and right-angled triangles, they were able to use the geometric relationships to locate hidden objects. The Munduruku did as well as American children on the same test.[7]

In another experiment, two researchers from Duke University determined that infants only seven months old grasped certain numerical concepts. The experimenters showed the infants videos of adults and at the same time played them recordings of adults speaking. The infants displayed a clear preference for watching the group of adults that matched the number of people they could hear speaking. This doesn't mean that babies can count, but at this preverbal level they grasp number sufficiently to be able to match it in the visual and the auditory domains.[8] The choice of adults and voices, experimenters point out, was not arbitrary. Not even children who are much older can perform in the same way if they are asked to match objects that matter less to them or that are less obviously related, like drumbeats and black dots. The infants' natural mental abilities are shaped by their environment. They are much smarter than we imagined, but their intelligence doesn't get expressed as abstract, computational efficiency; it's all about being human.

Many of the animals that demonstrate complicated thinking turn out to have a fair bit in common with one another and with us. Even though many of them are not that closely related to humans, they share many traits that seem as important as DNA. Hyenas, whales, elephants, humans, baboons, crows, and parrots all have long lives, extended periods of childhood, complicated systems of communication, and their societies are made up of individuals with distinct roles and relationships.

Accounting for the connection between phenomena like individuality and cognition is a fairly recent development. "In most studies of long-lived animals with elaborate social systems, the individual is extremely important because they have extremely varied experiences," said Betty's researcher Alex Kacelnik.

This is a familiar enough idea when we apply it to humans, who are pleased to take the performance of our best and brightest as evidence of our species' abilities. If you went to the Metropolitan Museum of Art to look at the Picassos, you wouldn't treat the art as just the work of one individual in highly special circumstances, but would likely examine it as an expression of what it means to be human. "We have different standards," Kacelnik said. "If a chess master says that he uses some unconscious process to learn what the next set of possible moves is, we call that inspiration and cognition. But say that you were to train an animal to play chess and you reward it for making appropriate moves in particular configurations of the board, you would not call that cognition. You would say that the animal has used trial and error. But you would be observing the same thing." Exploring the social complexity of an animal's life involves treating individual acts as part of the genius of the species rather than as exceptions to it.

Katy Payne and her assistant Melissa Groo at Cornell's Bioacoustics Research Program investigated elephant social complexity. Groo screened a video of a young female elephant calf they call Elodie, taken at the Dzanga-Sangha National Park in the Central African Republic. Elodie's antics took place in a *bai,* a muddy clearing in the middle of a forest, the elephant equivalent of a village square. Different families, each led by a matriarch, visit the *bai* over the course of a day, and at any one time up to eighty elephants might be scattered about. The young elephants play while the adults flap ears and rumble and thunder at one another. The elephants spend a lot of time using their feet and trunks to construct mud wells, holes that are a few feet in diameter. They stand in them and eat mineral-rich mud from the bottom. Generally, the dominant individuals (typically large adult males) occupy the best wells, while less dominant individuals stand around nearby waiting for a chance to slip in.

In the video Elodie enters the frame from the left. She is a tiny thing, trotting on huge feet, and she heads for a hole ruled by an enormous male. Given her size and sex, Elodie should be last in line for access to the well, but she walks in and plunges her trunk straight down next to the male's trunk; she is almost standing on her head—fat, round rump thrust up into the air and the rest of her not visible over the rim. Lamar, the male, is momentarily baffled by the interloper, so he lets her in. But quickly he recovers himself and pokes her in the butt with his tusks. Elodie screams and scoots out, and her mother, as always hovering by, jolts forward in response. "She is a nervous wreck," says Groo. But Elodie is not. Within a minute she sidles back up to Lamar and squirms into the hole again.

"You never see Elodie's behavior in other juveniles," Groo said. "It is a unique strategy." The ability to accommodate individualistic behavior like Elodie's within a group is an indicator of intelligence. It means that for elephants, as for humans, society operates according to a layered set of rules—on one level there are expected modes of behavior, yet on another level rules can be broken. This kind of flexibility requires a mental agility that would not be necessary in a social system based on a rigid behavioral pattern.

Lamar eventually tires of Elodie's intrusion and walks away, leaving the pit to her. But another male decides it is now his turn. Elodie's mother tries to stand in the way of her daughter's competitor and ward him off, but she is subordinate to him and quickly backs down. The male moves in on Elodie. He is not as big as Lamar, but he still towers over the baby elephant. Elodie will not budge, however, and shortly he yields and walks away.

Like crows, elephants are biologically distant from humans, yet like us they live long lives in structured societies where "childhood" is an extended period of learning out of which individualistic behavior emerges. The social demands of elephant society are intense. They include, Payne explained, growing up in a crowded community with members that change and develop over the years. For males, it means living in a very vocal, collaborative female society for their first twelve to fifteen years and then moving into a more silent, solitary, competitive existence. In their new world they make temporary associations and coalitions with other males, and they rise and fall in dominance as they go in and out of musth (heat). Like humans, female elephants live years past their reproductive stage. This means, Payne said, that elephant society is more sophisticated than societies in which the members do not live long, because the elders can impart their wisdom. Older females pass on social learning, like how to interact with hundreds of other familiar elephants, and also practical information, like where the best water hole or fruit tree can be found. This requires memory, knowledge, and the ability to learn that knowledge.[9]

Other researchers have commented on the sophisticated ways that members of animal groups such as these relate to one another. Frans de Waal calls the set of rules and relationships found in such complicated groups social syntax. Ray Jackendoff agrees there is a parallel to be drawn between the role of syntax in language and in social situations: "If you look at what the other primates are doing, you have to attribute some concepts to them. Not all of them by any means, but tracing who's related to whom and therefore who one is entitled to commit aggression against, these kinds of things require combinatorial structure, and they suggest that the meaning was around before the language." (See chapter 9 for more on the mental platform for syntax.)

The more we learn about what's going on in the heads of other animals, the more we realize that many different species have a lot to think about and their ways of thinking are quite sophisticated. Despite centuries of believing otherwise, we now know that it's possible to have a complex inner and social life without syntax and words.[10] Most significantly at this stage of language evolution research, the overwhelming accumulation of evidence for animal cognition resets the parameters of the problem—there can be no more easy assumptions about human uniqueness or the special status of our mental lives.

Researchers differ in how much they think our mental platform interacts with language, though most agree it has to have some role. At the most general level, examining the thinking of a broad range of species suggests how common certain types of cognition are among many animals. Narrowing the focus and looking at animals that live similar lives to ours or that are genetically closely related to us helps us consider what the mental life of our ancestors on the cusp of modern language might have been like. Based on the abilities of the chimpanzees, dolphins, parrots, and even crows described in this chapter, we can assume that their thought processes were already fairly complicated.

What does language bring to the mix? Ray Jackendoff, a linguist at Tufts, who fondly remembers the champagne atmosphere when he was a student of Chomsky's generative linguistics in the 1960s, argues that when you introduce language into the well-developed mental platform of pre-linguistic hominids, you get profound ramifications of thought, material culture, and social structure. "Language does help us think better," he said. "It doesn't enable us to move from zero to actual thought. Monkeys do have thoughts, and *you have to have something to say* before there is something adaptive in saying it."

Given the sea change in the way animal thought is viewed, Jackendoff outlined four logical possibilities for thinking about language evolution. First, some things that are necessary to language must have undergone no change at all from our pre-linguistic ancestors. Lungs and the basic auditory system belong in this group. Second, certain traits have appeared only in the human lineage, are relatively new, and are necessary for language but also serve a larger function. This group includes phenomena like pointing and the ability to imitate. Third, there are probably aspects of language that only humans have and that are used exclusively for language but are based on some alteration of a shared primate trait, like the shape of the vocal tract. Fourth, parts of language may be used exclusively for language and arise from a trait that is completely new and unprecedented in the lineage we share with other primates.

It is possible, Jackendoff acknowledged, that nothing fits in the third and fourth categories, and that language could be accounted for by traits and abilities that exist only in the first two. If this were the case, human language could be made up entirely of ingredients that are neither unique to our species or to language. Jackendoff, among others, doubts that this is the case. For example, he differentiates a number of abilities that seem to rely on conceptual systems that build on distinctions that can be made only in language. While these have not yet been studied extensively in primates (allowing us to rule them out as belonging to nonlinguistic cognition), they offer a good place to look for language-dependent cognition.

In a paper he co-wrote with Steven Pinker, Jackendoff described many ways of thinking that are not possible without language. These include fatherhood, moral concepts, tools made of three parts or more, ideas and systems of thought like the supernatural and formal and folk science, and kinship systems that make complicated distinctions like cross-cousins (mother's brother's child, father's sister's child) and parallel-cousins (mother's sister's child, father's brother's child).[11]

What about language and the concept of time? As with most other animal cognition research, we are just beginning to get a handle on how animals may think about time, whether consciously or subconsciously. Only recently we believed that animals lived forever in the present, unable to think about the future. But in 2006 Nicholas Mulcahy and Josep Call showed that orangutans and bonobos could plan for a future event. In a number of experiments Mulcahy and Call demonstrated that both kinds of animals were able to select from a range of tools the appropriate instrument for getting food out of a specially constructed device, *even though they wouldn't have access to the device for up to fourteen hours*. This series of experiments is the first to show that nonhuman apes can plan for a later need. Because our common ancestor with orangutans lived earlier than fourteen million years ago, Mulcahy and Call suggest the precursor for mental time travel is at least this old.[12]

Mulcahy and Call demonstrated how the concept of the future and future needs is not specific to humans. But what about our most complicated concepts of time? It's probably not possible to learn the way we carve up time without language, wrote Jackendoff and Pinker: "The notion of a week depends on counting time periods that cannot be perceived all at once; we doubt that such a concept could be developed or learned without the mediation of language."[13] Not only are ideas like a week reliant on the medium of language, but, Jackendoff and Pinker suggested, "more striking is the possibility that numbers themselves are parasitic on language—that they depend on learning the sequence of number words, the syntax of number phrases or both."[14]

A new generation of experimenters has begun to engage in earnest with the ways language, ideas, and thinking may interact. Gary Lupyan, a Ph.D. student at Carnegie Mellon University, studying under Jay McClelland (one of the founding fathers of connectionism), believes that language may shape cognition: "The idea that language affects thought has a great deal of intuitive support. We feel that we think in language and think differently in different languages. Languages around the world vary to an enormous degree, and so it would seem people speaking these languages ought to categorize and think about the world differently. Language seems to embed itself in so many aspects of our everyday cognition that we must start considering how language has altered the functioning of cognitive mechanisms we share with other mammals."

The question of whether language can affect the way we see or think about the world has long been controversial in mainstream linguistics. Edward Sapir and Benjamin Lee Whorf, two linguists working in the early part of the last century, first popularized the notion that a specific language can shape thought in a particular way. But in the Chomskyan era their theory fell out of favor. It was assumed instead that thought is structured by universal grammar, the core set of linguistic principles that all humans share. If this were true, then any effect of language upon thought would be the same for all people, regardless of which language they speak.

Either way, intuition is not sufficient for making assumptions about language and thought. Now researchers are subjecting Whorfian ideas to experimental tests, like that by Lera Boroditsky, a psychology professor at Stanford. As Lupyan described it:

> Boroditsky looked at speakers of Indonesian, a language that does not require tense marking. For example, an Indonesian speaker might say "I go," and it could mean going yesterday, today, or tomorrow. In addition, while not requiring speakers to mark tense, Indonesian does require speakers to provide information about the actor, such as relative

age. Boroditsky tested the Indonesian speakers' memory for different scenes, like that of a picture of a boy about to kick a ball, a picture of a boy kicking a ball, and a picture of a boy having kicked a ball. She found that English speakers had better memory for the tense, while Indonesian speakers had better memory for who performed the action.

We are finding that influences of language seem to extend into areas previously thought to be too low-level to be affected by it. I've found that the ability to mentally rotate objects seems to be affected by whether we have a name for the object that's being rotated. Language also changes how we remember colors and even actually see colors.

Research on color and language has a long history in psychology and linguistics because different languages divide the color spectrum differently. For example, among the many, many colors that it labels, the English language distinguishes blue from green, while many languages make no such distinction. In the past some studies have found that the way a particular language divides color can shape the way color is perceived, while others have found the opposite. The general consensus until now has been that different color labeling systems probably do not affect the color perception of individuals.

However, in 2006 Aubrey Gilbert and colleagues announced in the *Proceedings of the National Academy of Sciences* (PNAS) that the way a speaker's language distinguishes color does affect the way he or she sees it. The nature of their experiment had the additional benefit of providing a clue as to why previous experimental results have been so contradictory. Apparently, the way you see color depends on what side of the brain you are using.[15]

Gilbert and colleagues hypothesized that if language is dominant on the left side of the brain, it should impact the perception of input in the right visual field. (The left side of the brain controls the right visual field, and vice versa.) They showed subjects colors with color words, and found that subjects were able to make faster judgments about colors and color categories in the right visual field when the color and the word matched. If there was a conflict between the color and the word on this side, they were slowed down in their responses. The wrong word interfered with their ability to decide what the color was. When they were asked to make a judgment about colors and words in the left visual field, using the non-language-dominant side of the brain, they weren't affected at all.

It's not clear yet whether the language of color affects the way an individual physically perceives color in the world or whether the influence of language kicks in after some basic perception has taken place. Nevertheless, Gilbert's experiments show that linguistic categories affect thought. Lupyan said, "What we are now learning is that besides communicating information, language seems to alter how the brain processes it. Individuals, like stroke patients, who suffer from aphasia—a condition characterized by varying degrees of language loss—do not just find it more difficult to communicate; they also find it more difficult to categorize, remember, and organize information. This is evidence that language is playing a role in these cognitive tasks."

In his own research Lupyan addresses the question of how language in general, rather than specific languages, changes the way we perform cognitive tasks. He devised an experiment to tease out some of the ways words might affect how we think. Lupyan used a set of odd-looking clay creatures with prominent heads and strange pointy limbs, which he called aliens. His aliens fell naturally into two groups. In one, the creatures' heads were fairly smooth, and in the other their heads were somewhat lumpy and misshapen. Crucially, the differences were subtle and not easy to articulate. He then told two groups of students that some of the aliens were friendly and some were not. The students' task was to decide which was which, and then to assign them to separate groups. After they made each choice, students got feedback about whether they guessed right or wrong, meaning that as they went through the task, they basically learned that smooth heads were friendly and lumpy heads were not.

Lupyan added a little piece of information to one of the test groups. After the members of the group found out whether their choice had been right or wrong, they were also shown a word. Lupyan told them that previous subjects had found it helpful to label the friendly and unfriendly aliens, calling the friendly ones "leebish" and the unfriendly ones "grecious" (or vice versa). He found that even though both groups eventually learned the difference between the aliens with equal success, the group that had words to label them learned to distinguish them much faster than the non-word group. He concluded that language, specifically the act of naming something with a word, helps categorize.

"Separating language and thought is hard," Lupyan acknowledged. "But it is precisely because of this that we have to start thinking of them as not separate things, but as a system. As language is learned, it alters how we process information. Just as when we learn to identify a face with a name, it alters how we treat a face—it's not just a face, it's my friend Mike—so learning language results in our automatic labeling of objects, actions, sounds, and even more abstract categories like emotions. This labeling categorizes the item and links it to other instances of the category."

Language not only boosts cognition but can help or hinder thought, depending on the task in question. In 1990 Jonathan Schooler at the University of Pittsburgh demonstrated that when people were shown a face in a mock crime videotape and asked to write a description of it, they were worse at picking that face out of a subsequent lineup than people who hadn't written their impressions down. "This makes sense," said Lupyan, "if we think of linguistic descriptions as forcing us to think in categories. Writing 'he had brown hair' can impair later identification because 'brown' refers to a category and not a particular color."

Other language and thought experiments have looked at how we process number. Because of researchers like Wynn, Hauser, and their colleagues, we know that certain aspects of number ability do not depend on language, as some animals can think numerically, and children and adults use various number and geometric concepts independently of language. Nevertheless, a recent experiment suggests that some numerical concepts are difficult to understand if they don't exist in the language you speak.

Peter Gordon of Columbia University has studied the Pirahã tribe, which lives along the Maici River in Brazil. The Pirahã are known to the scholarly community because of the years of fieldwork carried out by the linguists Keren and Daniel Everett, the latter now professor of linguistics and anthropology, and chair of Languages, Literatures, and Cultures, at Illinois State University. There are only about two hundred Pirahã, who live in groups of ten to twenty and maintain a hunter-gatherer lifestyle, re-sisting assimilation into mainstream culture. They are completely mono-lingual and only occasionally communicate in a primitive pidgin with outsiders. There is no precise number system in the Pirahã language, which relies instead on a "one-two-many" categorization, distinguishing merely between amounts that are not much and those that are larger. For example, *hoi* means "roughly one" or "small"; there is no word for a singu-lar amount. Spoken in a different tone, *hoi* can also mean "two," as distinct from "one." *Baagi* or *aibi* designates amounts that are a few or larger. (This kind of system is not uncommon in many languages of the world.)

Gordon carried out a series of experiments on Pirahã speakers that were designed to test their numerical abilities.[16] In one, he sat across from his subjects, with a stick dividing his side from theirs. He positioned a line of evenly spaced AA batteries on his side and asked the Pirahã to place a similar array of batteries on theirs, matching each of his with theirs in a one-to-one correspondence. With each successive repetition of the task Gordon made it harder and harder by asking his subjects to match clusters of nuts to the batteries, or match orthogonal lines of batteries, lines that were unevenly spaced, or lines on a drawing. He found that the Pirahã were successful with two to three objects but had much more difficulty with larger numbers from eight to ten, where their success rate dropped to zero. The exception to this result was the test that asked the Pirahã to match unevenly spaced clusters. Although they had trouble with between three and six objects, they were almost perfect in matching seven to ten. Gordon suggests this was because the uneven display essentially allowed the subjects to break the larger amount down into groups of two and three.

The Pirahã's numerical abilities were consistent with the way infants and certain animals can make relatively accurate estimations of small num-bers of objects—up to three—concluded Gordon. Beyond this, if a per-son's language does not contain a number system that labels quantities like four and five, he may not have the ability to identify or use these numbers. The underlying concept is that languages that contain terms for higher numbers basically teach the learner-speaker to count at this level.

As experimenters become more sophisticated in their methods, it's reasonable to imagine that the ways that thought is ramified by the complexities of language will become more apparent. In the meantime, the work of Gordon, Lupyan, and others suggests that words are not just convenient labels for things; rather, they are extremely powerful mental devices. And if there is one aspect of language that appears to be a uniquely human and relatively recent innovation, it has to be the sheer size of our vocabulary. It's thought that speakers can have a vocabulary of sixty thousand words. But how old are words, exactly? Do animals have them? And if they do, does that mean that words have been around longer than humans?

6. You have words

In the 1980s two researchers from the University of Pennsylvania, Robert Seyfarth and Dorothy Cheney, published some attention-grabbing data about the communication of African vervet monkeys.[1] The researchers confirmed a 1967 discovery that the monkeys made specific, wordlike warning calls in response to particular predators. In a vervet group, all the animals are consistently looking around, and in a group of ten to twenty individuals someone usually spots any nearby predators. When it does, it gives an alarm. If the monkey making the alarm call saw an eagle, it would make one kind of cry sound; if it saw a leopard, it would make another; and if it saw a snake, it would make a different sound altogether.

Not only did the vervets produce different cries, the rest of the group reacted differently to each type of signal. If the lookout monkey made an eagle call, then all the vervets would take up the cry, echoing the sentry, while running beneath the cover of trees. Being under foliage was the best hiding spot in case of an aerial attack. If the lookout's alarm cry indicated the sighting of a snake, the vervets would do the opposite, climbing up into the trees and repeating the call—Snake! Snake! Snake! Up off the ground, in this case, was the safest place to be. If the sentry monkey spotted a leopard, it would make the leopard cry, and the vervets would likewise leap into the trees, but now they would climb out onto the narrowest, most lightweight branches. These were the perfect place to be if a hungry leopard was prowling, because the lighter branches wouldn't support the weight of the predator if it followed them up into the tree. In 1967 the vervet behavior was only observed. Seyfarth and Cheney replicated the observation of the vervets' different responses in experiments using alarm call recordings.

Each type of warning cry was consistently the same sound. It was as if there were three words that had been agreed upon by the whole monkey community, in the same way humans agree upon the arbitrary words of each human language ("eagle" if you speak English; *aigle* if you are French).

There was great excitement at these findings, which suggested that we'd finally found evidence of an animal word that worked the same way a human word does. The last common ancestor of vervets and humans lived around thirty million years ago. Was it possible that all you needed to achieve the complexity of human language was a proliferation of words, some syntactic rules to make them all work together, and thirty million years? And did this mean that words preceded humans?

For a number of reasons, it turns out, the answer is probably no. But it is a gray kind of "no," and the reason the vervet cries are not satisfying candidates as animal words is not the most important thing about them.

Vervet alarm-calls-as-words had such appeal in the scientific community and the popular press in part because these animals are relatively close kin to humans. If you think of chimpanzees and bonobos as our brothers and sisters, and gorillas and orangutans as our aunts and uncles, then the vervets might be third or fourth cousins. Alarm calls from vervets were much easier to imagine as the antecedents of our language than if they had been coming from, say, a chicken.

But alarms calls are ubiquitous in the animal world. Monkeys have them. Ground squirrels have them. Meerkats have them. As recently as 2005, researchers in the journal *Science* discussed the complicated and clever alarm calls that chickadees make. And, yes, even chickens make alarm calls, distinguishing between terrestrial and aerial predators.

"Most birds," said Tecumseh Fitch, at the University of St. Andrews in Scotland, "have a sort of generalized alarm call and an aerial predator alarm call. It is by no means unusual in the animal kingdom to have at least two different kinds of alarm for two different types of threat. Ground squirrels have about eighteen calls, and meerkats have more alarm calls than vervets simply because they have more predators than vervets."

Even though humans are more closely related to vervets than vervets are to chickens, it appears that vervets and chickens have converged upon a common tactic for survival. The forces that led them both to this strategy are powerful, but alarm calls were probably not bequeathed to them from a common ancestor. In fact, the most important thing that they share with all the other alarm-call-making animals is that they are small and delicious. Fitch explained: "The things that have alarm calls are little tiny guys who get eaten by lots of things, and the common ancestor of chimps and humans wasn't in that category. Humans don't have alarm calls, and apes don't have alarm calls. It's not that they don't have threats, but they don't have all these different threats where it pays to be able to refer very rapidly to aerial threat versus ground threat. Whether you're the Snickers bar of the Sahara or the Snickers bar of South Dakota, you're going to evolve alarm calls."

Fitch discusses the evolution of communication with enormous energy. He was named for his great-great-great-grandfather William Tecumseh Sherman, who ended the Civil War with his famous march from Atlanta to the sea. (His ancestor, in turn, was named for the Shawnee Indian leader Tecumseh, who traveled up and down the East Coast, uniting tribes in opposition to the spread of the United States.) Fitch himself occupies a unique spot in the new field of language evolution. He studied under Lieberman, writing his Ph.D. thesis on the evolution of speech. And in 2002 he collaborated with Noam Chomsky on the first paper Chomsky wrote about the evolution of language.

"Alarm calls seem to be a prime candidate for language evolution," Fitch said, "but they are not." The calls aren't like human words, because they are genetically preprogrammed: animals will produce them even when raised in isolation. "What vervets have is the ability to communicate a very limited set of meanings," he explained, "and because that's genetically determined, there's no way other than genetic modification to add new units into the system. Each call type has to evolve over Darwinian time, and you can't evolve limitless meaning, as you have in human language, in Darwinian time."

What would English speakers look like if they inherited sixty thousand words genetically? It's hard to imagine. Babies would presumably be able to talk from birth, and they'd have an enormous memory capacity. Most animals that have a lot of information genetically coded are born looking fairly well developed. Our nine months of pregnancy might be considerably longer. New words—and the ideas or innovations they represent—would have to propagate through the species genetically, so adding a single word or idea like "wheel" or "fire" or "cooked meat" would take a few thousand years. Science, art, and McDonald's would just never get off the ground.

If alarm calls aren't words, then what are they? "They're not words in the same sense of language," Fitch explained. "They're more like laughter and crying, which are also calls that are innate. You don't need to hear your mother crying to learn how to cry. Deaf children make these sounds, too." And as you grow, you learn that when you laugh, people nearby can safely assume that something you find amusing has occurred. If you burst into tears, they can likewise guess that something you find upsetting has happened. "No one has to have any recourse to words to make these sounds or to interpret them," Fitch said.

We don't know exactly how these calls evolved, but it's not hard to imagine that if you were a vervet monkey with a tendency to laugh hysterically and run up a tree every time you saw a leopard heading your way, then you and your troop might end up more likely to survive and to reproduce. When you did reproduce, you'd pass on that genetic predisposition to at least some of your children. The stoic monkey would be a dead monkey.

Instead of seeing alarm calls as a primitive form of language, we should look at them as a communication device that many animals share. Across a wide swath of life, animals as genetically distant as birds (famously descended from dinosaurs) and mammals have evolved distinct units of sound that act as pointers to things in the real world.

It could even be argued that human calls—laughter and crying, which certainly intersect closely with language—are a degenerate form of the alarm calls of prey species. When people hear you laugh, they know you are laughing at something, but don't necessarily know anything else about it. When vervets make the eagle call, other vervets know that something scary and aerial is headed their way and that they should look up, as opposed to around or down on the ground. In this regard, they make more reliable and specific inferences than we do.

Some researchers still think it's possible that alarm calls are a kind of protoword—that we somehow broke the link between the vocal token and the DNA, retaining the ability to use freely a sound token to refer to things in the world. There is some interesting neurological evidence for this possibility. Chris Code, a research fellow in the School of Psychology at the University of Exeter, points out that it is possible neurobiologically to separate swearwords from other words in language. Swearing actually uses parts of the brain that support language and also parts of the brain that are used when laughing and crying. Often people with severe brain damage remain able to swear even when they are unable to produce other language. Perhaps swearing is the remnant of an evolutionary step at which cries were some mix of automatic and voluntary articulation. While the possibility cannot be ruled out altogether, the safest conclusion at this stage is that alarm calls are probably not the antecedents of words.

The vervet story invokes many of our muddled ideas about animal communication and how it compares with human communication. A few themes crop up again and again. There is the notion that animal vocalizations are just gibberish, the opposite of language, and much like what we produce if we cry out nonverbally—informationless sound that provides a crude guide to an emotional state. And there is the contrary idea that animals use a code to communicate with one another, as we do, but we just haven't cracked it yet. Both these approaches assume that animal communication will be recognizable in the terms we use to understand our own language, that it has words, or it doesn't. It has syntax, or it doesn't. It is full of meaning, or has no meaning and no reference whatsoever.

Another suggestion is that other animals may communicate using degenerate, primitive tokens of our own language. This is part of the broad assumption that humans are intellectually and communicatively superior to all other animals. The vervet alarm calls seemed to fit nicely into this concept: the monkeys had words, but just three of them.

It is true that the complexity of human language is without parallel. It enables us to connect with one another in a virtual world and together invent agriculture, construct buildings, send airplanes through the sound barrier, and shoot satellites into space. But assuming that the most salient thing about human language is that it is *the* superior form of animal communication doesn't get you very far. It doesn't tell you what parts of language may have been positively selected. And it can't tell you about how language evolved. It implies that anything can be expressed by human language, when we don't know if this is in fact the case.

This approach also implies that human language is the communication tool par excellence, as opposed to a communication tool that developed in a certain niche. But assuming that language is the best possible communication tool is a little like saying that the human brain is quite simply superior to all other brains, as if our brain was an all-purpose machine rather than a device that does some jobs very well and others less so. Such evaluations take the trait, like language, or the organ, like the brain, out of the context of the body and the niche, as if evolution acted independently of the needs of the organism in its environment.

Indeed, saying that the only important thing about language is what it does better than other communication systems is as nonspecific and unhelpful as saying that humans are the most intelligent species—which is itself like saying humans are the best-looking species. If you understand this sentence, then you already belong to the species that agrees with the sentiment.

What matters about the alarm calls is surprisingly obvious but until now has rarely been commented on: when vervets and chickadees and chickens make their alarm calls, they are connecting a particular sound to a referent in the world. Whether the animal arrived at this behavior genetically, somewhat genetically, or not, it appears that it is a widespread, easily evolved, and useful trait.

"Every species where researchers have tried—and that includes dogs, dolphins, parrots, and chimps—can link sound and a reference," said Fitch. "I don't think this is some sort of special human ability. It's a pretty general ability. What else is your brain for? If your brain can't link two stimuli in the world, one which is visual and the other which is auditory, then what good is it? I wouldn't be surprised if fish could do this, but no one has really tried to see if they can."

So the act of hearing a particular sound and making meaning out of it is not particularly human; it's ancient. Animals like vervets use the connection between a sound and a visual signal in one way, and humans have built on this ability in another way, using it as a platform for human language.

In order to progress, science has to focus closely on some areas to the exclusion of others, which sometimes means that the most obvious facts of our daily lives are ignored. For instance, humans communicate with dogs. That observation is so mundane it hardly deserves mention, but this ability is relevant to understanding what evolved in order for us to evolve language and how the platform for understanding a word is ancient. Philip Lieberman spoke about the relevance of this ability in dogs in his 2000 book, *Human Language and Our Reptilian Brain*, but only in the last few years have other researchers begun to actively investigate it.

In 2002 a team of researchers at the Max Planck Institute in Germany showed that Rico, a Border collie, knew the meaning of hundreds of words. Not only was Rico able to go into another room and retrieve an object he had been asked for (choosing it from a selection of possibilities); he was able to infer the meaning of words he'd never heard before. For example, Rico was asked by the researchers, who used a word he didn't know, to go and retrieve an unfamiliar item. When he went into the next room to look for it, only one object in the set of possible things to retrieve was one he had never seen before. Because Rico knew the words for all the other objects, he picked up the novel one, assuming it was what the experimenters were asking for. Rico, obviously, does not have human language. Instead he is using an ancient, more general skill that preceded language by millions of years. It's at this level that humans and dogs communicate with each other, as with an animal like Alex, the African gray parrot.

Other animals appear to have built on the ability to make meaning from the connection between a sound and a referent, as with human words. Dolphins use echolocation clicks, "burst-pulse sounds," and different types of whistles. "Signature whistles" are so named because it appears that dolphins name themselves.[2] These beasts reproduce a distinct, individual sound that develops in their first year of life whenever they meet another dolphin. It's always the same, and always distinct from any other dolphin's whistle. There is even some evidence that dolphins will exchange their signature whistles when separating. In 2006 a team of researchers led by Vincent Janik at the University of St. Andrews in Scotland found that wild dolphins recognized that a signature whistle referred to a particular dolphin even when its voice was completely distorted.[3]

Elephants also appear to use sounds like words. Katy Payne, lead researcher on the Elephant Listening Project at Cornell University's Bioacoustics Research Program (now retired), and Joyce Poole, scientific director of the Amboseli Elephant Research Project in Kenya and another longtime elephant researcher, began an elephant dictionary study. The goal was to describe the way that individual elephants produce distinct sounds for various purposes, like greeting a fellow member of the clan they haven't seen in a while. Dolphins and elephants don't have words as we do, but both of these socially complex species have instead hit upon some of the same tactics to communicate.[4]

Chimpanzee pant hoots are another interesting wordlike call. The pant hoots are very loud cries that are most often used to communicate over distances. Their function seems at least in part to be to rally support and keep individuals in a group together. Pant hoots also differ between individuals and between different chimpanzee groups. Chimpanzees appear to be able to pick and choose which ones to use. They are somewhat like dolphin signature whistles because they seem to have an internal structure and are uttered in various situations, such as resting, feeding, and during travel and display. This suggests that chimpanzees have some ability to choose meaning, as well as use structure.

Klaus Zuberbühler and Katie Slocombe of the University of St. Andrews recently investigated the ability of chimpanzees to make humanlike reference in an experiment at Edinburgh Zoo. The researchers monitored the chimps and found that they issued distinctly different cries in response to finding different kinds of food. When the chimpanzees came across highly valued food, like bread, they made high-pitched grunts. When they came across food that was less appealing, like an apple, their grunts were low-pitched. Zuberbühler and Slocombe demonstrated not only that the chimpanzees were making distinctions in the way they vocalized about their food but that other chimpanzees seemed to understand the meaning of the different grunts. When they played recordings of the various food grunts to chimpanzees, the listeners would search for the given food in the place where it was usually found in their pen for a longer time and with more effort than in other spots where different food might be found. The chimpanzees also searched longer if the cry signaled a particularly prized piece of food.

These findings suggest that our closest relatives have built upon the sound-referent connection to communicate distinctions to one another in a similar way that we do. There is more than just genetics involved with pant hoots, as well as signature whistles, and time and more research will help tell us the ways in which these sounds and meanings resemble human words. Certainly, it looks as if the voluntary production of sounds that are meaningful to another creature is not a uniquely human ability.

Another question raised by the vervet studies is whether vervets intend in any conscious sense to communicate. Seyfarth and Cheney have carried out experiments on captive vervets in which they exposed adult females to a "predator" when they were either with a juvenile offspring or with an unrelated juvenile. The females gave many more alarm calls in the former case than in the latter. They also observed, in the field, one instance when an isolated vervet was being pursued in a tree by a leopard. The vervet gave no alarm calls, suggesting that the animals can withhold calls when no other monkey is around.

As for the interpretation of the alarm cries: the monkeys learn what each cry means only through experience. (Some researchers argue that the cries induce only an emotional, not a cognitive, response in other monkeys, and that's why they run. But as Seyfarth and Cheney point out, the fact that the responses are obviously emotional does not rule out the possibility that information in them is interpreted as well. Wouldn't you also feel panic if someone screamed, "Snake!"?)

The ability to interpret another's animal utterance is so universal that even animals of different species can understand the cries that other creatures make. Seyfarth and Cheney have observed that predators that hear the alarm call of their prey often give up the hunt at the sound—they know they've been seen.

Said Fitch, "What I think is interesting and surprising and we didn't know twenty years ago is that animals have an asymmetry between perception and production. This appears to be one of the key differences in being able to communicate with words and not."

While other mammals appear to be very good at making meaning from sound-plus-reference combinations, they don't necessarily produce new sounds in connection with new objects in the way we do. "The intuition is that if you can see something you must be able to produce a word for it," Fitch explained, "but that's where the data is completely clear—it's not so. Dogs can bark, but they do not create new barks to correspond to new sounds, and chimps can scream, and they can even withhold their screams in certain contexts, but they can't freely create new screams to correspond to new things."

Clearly there are varied ways that ancient capacities are used by different species, but being able to both understand and produce words is one of our special talents. Over time, we have produced hundreds of thousands of words, and there is little evidence that animals naturally produce many wordlike tokens at all. Individually we learn tens of thousands of words in a lifetime, and if we want, we can make up as many as we like. Language is in constant flux, so regardless of our own individual contribution to language change, words do inevitably become altered over hundreds of years. Rico and Alex and many other animals are able to comprehend that new sounds can refer to new objects, but they are not even remotely as adept at inventing words themselves.

Human words are much more than just links between sound and reference in the world. Indeed, reference is not the half of it. A word is an arbitrary association between sound and meaning. There is nothing in the sound of a word that tells you what it means or what it does—you must learn this as a child.[5] Whenever you hear a word, you know what (if anything) it refers to. You know that some words stand alone, like "hello," "ouch," and "yes,"[6] and that others can join together to create larger words, like "heretofore" and "bedroom."

You know that a word is a noun or a verb or another part of speech. If it's a noun, you know how to make it plural or singular. If it's a preposition, you understand that it relates two nouns together in space or time. If it's a verb, you know how to render it in a handful of different tenses, and you know what nouns will make sense with it and which ones will be nonsense. You know that some verbs have to have agents, like "killed." All of this information about a word is specific to language. Of course, you know that "table" refers to a table, and on this level learning a word and learning an object may not be dissimilar processes. But all of the information about the way a word combines with others in language is internal to the language that you learn.

You may not consider most of this consciously, but when you learned language, you internalized all of this information, and when you hear any word, you use this knowledge in the way you process it. You know all sorts of things about just the sound of the word. You know what other words will rhyme with it. You know which words start out with the same series of sounds, even though they end differently.

A child's ability to learn many words is so completely different from anything observed in other species that many researchers propose that some neural mechanism must be especially dedicated to this acquisition of linguistic knowledge.

Beyond the basic link between an unanalyzed sound and a simple reference in the world, words are clusters of complex knowledge about sound, grammar, and meaning. Human words don't exist by themselves. They are points in a series of intersecting systems, and when you hear or produce a word, all these systems come into play. Recent research has shown that when children acquire words, they are not just creating a multidimensional connection between different kinds of linguistic and nonlinguistic knowledge based on a platform of sound and meaning. The essential scaffold for word learning is more complicated than that. As well as a connection between two domains, such as the aural and the visual, there is a very important connection between speaking words and gesturing meaning.

7. You have gestures

Picture the house in which you grew up. Think about the rooms, the hallways, the stairs, visualize where they all are. Where was the front door? The back door? What color was the roof? Did you have wall-to-wall carpeting or were rugs spread all over the place? If you turned now and attempted to describe the house to someone nearby, it's highly likely that you'd gesture as you spoke. In fact, even if you just imagine a person and then describe the house aloud to her, you'll probably gesture as well. Gesture experts say that it is almost impossible to talk about space without gesturing. Gesture is spontaneous, and as integral to individual expression as it is to communication. Even though you probably won't gesture as much if you are talking on the phone, you will still wave your arms about. Blind people gesture when they speak in the same way that seeing people do.

Gesture may be integral to human expression, but it is not uniquely human. At the Gestural Communication in Nonhuman and Human Primates conference in 2004, Mike Tomasello of the Max Planck Institute in Leipzig, Germany, and his associates presented a huge compilation of gestures that they had observed in monkeys, gibbons, gorillas, chimpanzees, bonobos, and orangutans. Many of them had been observed at the spectacular ape exhibit at the Leipzig city zoo, where a leafy path leads to the center of a big ring. Radiating out from the central space are walks that divide all the great ape species from one another. In one section are the gorillas, sitting impassively. In another are the bonobos—only three of them, a reflection of their dwindling numbers worldwide. In the third section are the orangutans. The male sits near the viewing window looking profoundly deflated, while his orange cage mate hangs upside down from a tree stump and stretches. In the fourth section are the chimpanzees. There are more than a dozen chimps in the compound, and they make a lively community.

Some recline sensuously, others fly through the air on ropes or trunks. Some busily work at boxes, inserting sticks into various holes. The exhibit is climate-controlled; it feels like a light summer day. Tomasello has a number of testing rooms installed at the zoo for his various experiments.

Gestures play a large role in primate communication, Tomasello explained, and as is the case with humans, these gestures are learned, flexible, and under voluntary control. Most primates, humans included, gesture communicatively with their right hands, suggesting that the dominance of one side of the brain for vocal and gestural communication could be as old as thirty million years. Just as with human gestures, ape gestures can involve touch, noise, or vision. Apes wait until they have the attention of another ape before making visual gestures, and often if their visual or auditory gestures are unacknowledged, they will go over to the ape they want to communicate with and make some kind of touching gesture instead. Apes also repeat gestures that don't get the desired response. Like human gestures, ape gestures seem to be holistic: a series of gestures doesn't break down cleanly into meaningful components. Moreover, a set of different gestures may mean just one thing, while a single gesture may be used to convey many meanings.

Tomasello and his group divide ape gestures into two types: attention getters and intention movements. Attention getters, said Tomasello, slapping the podium, do just what they say—they call attention to the ape making the gesture. Chimpanzees will hit the ground, clap their hands, and stamp their feet for this purpose. They also lay their arms on other chimps, tug on their hair, or poke them. Once the observer pays attention to the gesturing ape, said Tomasello, what is required becomes clear. To illustrate this, Tomasello showed a video of a chimpanzee who walks over to another chimp and starts jumping up and down on the spot. When the second chimp finally notices the display, the first one turns around and sits down. The message is obvious—groom me, and that's what the second ape starts to do.

Intention movements are the beginnings of an actual movement, like a raised fist to indicate a threat in humans.

The process by which these gestures evolve in individuals, Tomasello explained, goes like this: "I'm really doing something, you come to antici-pate it, I notice your anticipation so I only make the beginnings of the movement." Male chimpanzees, for example, make a penis-offer gesture to propose sex. They sit back on their haunches and repeatedly thrust their pelvis, pushing their erect penis in the direction of another chimpanzee. "In papers we call it the penis offer," Tomasello said. "Between ourselves, it's called 'dirty dancing.'"

Mimicking another intention movement, Tomasello rolled his arms over his head, like a chimp barrel-hitting a companion. The move is remi-niscent of the way that humans feint at each other to make a point without actually following through. Cats and dogs make a similar movement when they raise their paws and bat them, as if they are about to strike another animal, so the gesture is not restricted to primates. "It's typical mammalian play," Tomasello explained. "Remember," he said, invoking the tree of life, "it's not a ladder; it's a tree. It's not a ladder; it's a tree."

Another gesture researcher, Joanna Blake at York University in Canada, directly compared the gestures that infants make when they are learning language with the gestures made by apes, which have a lot in common. Both apes and children make a lot of request gestures—begging for food, raising their arms to be picked up and carried—and they extend their whole hand to point. Children and apes likewise make the same gestures of protest, pushing someone away or turning away themselves while shaking their heads. They also emote in the same ways, stamping their feet, flap-ping their arms, and rocking, and when they want someone to do some-thing, both take a person's or an ape's hand and place it on the object to be manipulated, or they proffer objects that they want someone to manipu-late. Clearly there is a close family relationship between human and ape gesture, confirming that it is an ancient trait that precedes the existence of modern humans and of language.

Janette Wallis, who has been watching primates since she was an under-graduate at the University of Oklahoma, is drawn to the more subtle as-pects of primate communication. She used hidden cameras to capture evidence of a baboon gesture she calls the muzzle wipe—a quick pass

across the bridge of the nose with the hand. The muzzle wipe typically occurs in situations in which a baboon may be nervous or conflicted for some
reason. As with many human gestures, there's no evidence that the wipe is
intentional, but it's likely that other animals read it as a signal that reveals
information about the wiper.[1]

Wallis presented videos of the muzzle wipe at the Leipzig gesture conference. Although most early studies of baboons, she said, hardly mention
the gesture, her films showed baboons doing it in captivity and in the wild.
The gesture rarely lasts longer than a few seconds, so it is not easy to see,
yet once Wallis told the audience what to look for, the muzzle wipe was
clearly evident. Nervous baboons could be seen constantly putting their
hands to their faces in difficult situations. She noted that monkeys make a
similar move and that a chimpanzee will often put its wrist to its forehead
in similar contexts. Could this overlooked gesture be some kind of precursor to comparable gestures in humans? asked Wallis. Humans do put
their hand to their face when nervous, and indeed, as she pointed out, psy-
chiatrists and law enforcement officials often interpret a hand-to-face gesture as evidence of uncertainty or even deception.

Once Wallis convinced the audience that the muzzle wipe existed, she
showed a video of George H. W. Bush. The ex-president was speaking at a
press conference about his son the president of the United States. He discussed what was at the time headline news—George W. Bush's having
been arrested in his youth on a drunk-driving charge. "Unlike some," said
the older Bush in a tone of complete confidence, "he accepts responsibility." He then raised his hand to the bridge of his nose and scratched it.[2]

Only ten years ago researchers were unanimous in their agreement that
pointing was unique to humans. Even now many stand by that claim. In
fact, apes and many species of monkeys that are much more distantly related to humans do point as well, though they typically do so with their
whole hand.[3] (Scholars of gesture complain that pointing with the hand has
been treated as a second-class kind of pointing, even though it is common
in many human groups.) Usually, apes make this gesture only for humans,
not between themselves. They point at objects and alternate their gaze between the object that is pointed at and the human they are pointing for.

The animals learn how to point without explicit training, and simply pick it up from humans.

Although there is only one anecdotal report of a bonobo's pointing with its index finger in the wild, some apes have been shown to do so in captivity. William D. Hopkins, a researcher at the Yerkes National Primate Research Center at Emory University, and his colleague David Leavens, a professor of psychology at the University of Sussex, showed a videotape at the gesture conference of a chimpanzee pointing. In the video, Leavens is in a white lab coat and a surgical mask while a chimpanzee stands eating on the other side of a wall of wire mesh. When the ape drops some food through the mesh, it points its index finger through the wire to indicate the food and looks at Leavens, who picks it up and returns it. "I submit," Leavens said, "that there is a well-trained primate in this video, but it is not the chimpanzee."

At the Leipzig conference Tomasello was skeptical that apes could point and, if they did, that it actually meant anything. But he began to wonder about it and later said, "Many of the aspects of language that make it such a uniquely powerful form of human cognition and communication are already present in the humble act of pointing."

Tomasello had already established in previous experiments that apes know what other apes are seeing, and it was clear that they gesture easily and creatively for one another. More recent experiments have shown that chimpanzees will cooperate with one another in situations where collective help is needed (in order to get food, for example), and in quite simple tasks they'll also assist without the prospect of a reward—like picking up a dropped object and handing it to someone. While the Hopkins and Leavens video showed they are capable of pointing, why, Tomasello asked, do apes point only for humans and not one another? The answer he arrived at is both simple and far-reaching: it is because humans respond. Apes don't point referentially for other apes, because they will be ignored.

Human children learn to point at a very young age. Tomasello and his colleagues have videotaped many instances of children spontaneously pointing in a helpful manner. In one experimental setup, a very young child was placed on her mother's lap. Mother and child sat across a desk from a woman stapling papers together. The woman left the room for a moment, and while she was away a man entered, took the stapler, and placed it on a cupboard behind the desk. When the woman returned she made a great show of looking for the stapler. The infant watched her for a while, and then, unprompted, pointed to where the stapler had been moved so the woman could find it. In other examples, a child and adult played together until for some reason (the ball dropped, the toy fell) the game stopped. Without prompting, the child looked at the adult and pointed to the problem, clearly requesting that the game begin again. In other cases, the child pointed at an object or proffered it merely to show it to the adult in order to elicit a reaction.

Tomasello first started to consider how much this kind of shared, cooperative attention mattered at dinner in a restaurant one night. He was watching a mother and child play together. The mother blew a raspberry on the child's arm, then the roles were reversed, and the baby followed suit. Why did it happen this way? wondered Tomasello. Why did the child reciprocate the gesture rather than simply imitating the action on himself?

The answer, he believes, is that humans are particularly cooperative in the way they communicate.[4] Reciprocation is fundamental to the interactions of our species. Offering is not instinctive for humans, but is taught by parents to children, who learn it very easily. And crucially, we offer not only food and other objects but information and experiences as well. Children, says Tomasello, want you to look at what they are looking at and to emote in response. In many theories of evolution, human altruism is treated as an anomaly. But Tomasello thinks of it as an evolutionary strategy that has served us incredibly well.

Chimps don't spontaneously point in this fashion, and Tomasello believes it is due to a fundamental difference in the balance of cooperation and competition within the species. Chimpanzees lack the set of skills and motivations that underlie our pointing. Tomasello conducted an experiment with Brian Hare, then a doctoral student, in which two barrels were set up in a room. Food was placed in one, while the other was left empty. Hare stood on one side of the barrels as a chimpanzee entered the room. In one run-through, Hare pointed helpfully at the barrel with the food in it. But, said Tomasello, the chimpanzee would look at the finger, and then look at the barrel, and then look at the other barrel, and then it would choose completely randomly between them. It did not comprehend that Hare was being helpful and telling it where the food was located. In another run-through of the experiment, the chimpanzee would come into the room, and instead of pointing to the food, Hare would reach for the barrel, as if to grab it and the food in it. The chimpanzee understood this gesture without any problem, and it would head for the appropriate barrel. The movement Hare made was essentially the same in each case—a basic arm extension—but his intention was clearly cooperative in the first instance and competitive in the second.[5]

Tomasello and his colleagues' gesture work demonstrates both a continuum that connects human and ape communication and significant differences between them. In our evolutionary history some individuals must have been born with a greater inclination and ability to collaborate than our common ancestor with chimpanzees. These individuals were more successful and bred more offspring with those characteristics, Tomasello said. What we have evolved into now is a species for whom an experience means little if it's not shared. Chimpanzees took a different path. In their communication, there is never just plain showing, where the goal is simply to share attention. While they do share and collaborate and understand different kinds of intentions, they don't have communicative intentions. We do, said Tomasello, and it's in this shared space that the symbolic communication of language lies.

Tomasello's conclusions resonate deeply with observations made by Sue Savage-Rumbaugh. Before Kanzi, Savage-Rumbaugh worked with two apes called Sherman and Austin. The apes had successfully acquired many signs and used them effectively. There didn't seem to be anything odd about their language use until one day they were asked to talk to each other. What resulted was a sign-shouting match; neither ape was willing to listen. Language, wrote Savage-Rumbaugh, "coordinates behaviors between individuals by a complex process of exchanging behaviors that are punctuated by speech."[6]

At its most fundamental, language is an act of shared attention, and without the fundamentally *human* willingness to listen to what another person is saying, language would not work. Symbols like words, said Tomasello, are devices that coordinate attention, just as pointing does.

They presuppose a general give-and-take that chimpanzees don't seem to have. For this reason, Tomasello explained, "asking why only humans use language is like asking why only humans build skyscrapers, when the fact is that only humans, among primates, build freestanding shelters at all . . . At our current level of understanding, asking why apes do not have language may not be our most productive question. A much more productive question, and one that can currently lead us to much more interesting lines of empirical research, is asking why apes do not even point."

Whether you are human or another kind of ape, one of the ways that gesture becomes ritualized and communicative is in being passed on by learning. As humans, we observe a gesture, and then we reproduce it by imitation. Imitation is crucial to the learning process, and we are not the only imitators in the animal world. Lori Marino, one of the researchers who explored the ability of dolphins to recognize themselves in mirrors, said that "imitation is an everyday behavior with dolphins." They are very good at shadowing, imitation in real time. "If you make certain hand gestures in front of the tank in a captive facility, they will be able to follow your hand, even when you're moving your hand back and forth in different ways. They also seem able to pick up patterns very well and anticipate patterns, so if you set up a certain pattern going and then you stop, they seem to anticipate what the next step in the pattern is."

Frans de Waal speaks of the difficulties of measuring fleeting and ephemeral behaviors like imitation. "A lot of the cognition studies are on technical cognition, like: Can they count? How do they use tools? Do they understand gravity? Social intelligence is more difficult," he said.

Particular difficulties arise with imitation studies, as de Waal explained:

> What people do, for example, in these imitation studies is they put an experiment in front of the chimpanzee and they show how to do something, and then they see the chimp imitate. But I think imitation also requires that you identify with the person and that you like the person actually. If you look at humans who imitate, children who imitate, they imitate the people they know and they like, and they want to be like

Mom or they want to be like Dad or their big brother or whatever. They're not imitating a random person. It's very selective. I think the scientists who have failed to come up with these social learning tasks on chimpanzees, to some degree, have worked with the wrong paradigm. They put a human in front of the animal, which is already a different species, and the human may not have much of a relationship with them. I think we can only resolve these issues by focusing on behavior among animals themselves.

De Waal has been studying the ways that capuchins imitate one another. The experimenters train one capuchin to perform a task, and while other monkeys watch it, they attempt to determine if any imitation is taking place. De Waal is also probing the relevance of who gets imitated—if a capuchin is more likely to imitate its mother, for example, than an unrelated male.

Sue Savage-Rumbaugh's experiences with Kanzi back up de Waal's observation about laboratory experiments. She noted that Kanzi's mother, Matata, had two other children who never got the amount of attention from human caretakers that Kanzi did. She believes it was the significant relationships with humans in the period in which Kanzi was most sensitive to acquiring language that enabled him to pick it up.[7]

Other research suggests that imitation can be affected by who the original performer is. One recent study described the way a population of dolphins off the coast of western Australia passed on a tradition of tool use. These dolphins learned from adults in the pod to use sponges to forage on the ocean floor. But they didn't just acquire the skill from any of the adults: the tradition seemed to be passed down solely from mother to daughter.

The combination of gestural communication and imitation can be as powerful as vocal communication. In human hunter-gatherer groups, such as the Ngatatjara of western Australia and the northern Déné of the Canadian subarctic, the transmission of knowledge about the environment and how to survive in it is achieved by observation and experimentation rather than by verbal explanation. Moreover, studies have shown that a group learning how to manufacture a stone flake (such as those used by Stone Age societies) from a teacher who only gestured took no longer at the task than, and were as good at it as, a group in which the teacher gave precise verbal instructions on how to make the flake.[8]

In modern humans gestures come in a variety of types. There is here-and-now pointing (this book, right here!), action gestures (she picked it up with one hand!), abstract pointing (and another thing!), and metaphorical gestures that make symbolic reference to people, events, space, motion, action. Most gestures are initiated with the right hand. They typically occur slightly before or at the same time as speech.

Gestures that accompany speech typically amplify the meaning conveyed by the speaker. Sometimes, gesture communicates information that isn't explicitly stated in the verbal message it accompanies. For example, a speaker may move his fingers stepwise in a spiral while saying, "I ran all the way upstairs." The listener can infer that the staircase was spiral even though the fact was not stated.[9] While gesture doesn't break up into word-like segments, there are rules about the way gestures can be combined. And as obvious as the meaning of many gestures is when they are used by people while they are talking, listeners can usually guess at the meaning of a gesture without sound only 50 to 60 percent of the time. (Think about gesturing while saying, "I had a big ball" and "The guy had a huge hot dog.")

For a long time gesture was more or less ignored in linguistics, and elsewhere it received little attention. Researchers considered it paralinguistic, meaning that it was merely supplementary to language, perhaps useful in terms of emphasis but ultimately a secondary and unimportant phenomenon. People assumed that gesture was only for the benefit of the listener and justified removing it from serious consideration for the simple reason that it could be removed. It is possible, after all, to hear and understand someone even if you don't look at him. (In the same way, structure in language has been treated as separate from meaning, because you can go a long way analyzing both of them without reference to the other. Similarly, intonation has been largely ignored within Chomskyan linguistics.) The assumption was that because you could separate them in analysis, they worked independently in the body and they therefore evolved independently of each other. But even though you can discover much about speech and language without worrying about gesture, the fact is they usually occur together in the real world. Speech is disembodied only on the phone or radio, and in evolutionary time these types of communication have not been around very long.

Today, like the study of language evolution itself, the field of gesture studies is undergoing a small revolution. More and more people are engaging in experimental studies of gesture, and researchers are discovering how complicated and interesting it can be. Conference organizers in the last few years have been surprised at the number of scholars who want to attend meetings about gesture. This mini-boom is part of the general trend to reconsider what used to be called the epiphenomena of language. In a relatively short amount of time, researchers have shown that speech and gesture, as well as gesture and thought, interact as language is being learned and even after it has been fully acquired.

Traditionally, developmental psychologists thought that children gestured simply because they saw their parents do so. They believed that infants acquired language separate from any gesturing and in a predictable pattern. There was a one-word stage, followed by a two-word stage, and once a child crossed a critical threshold into a three-word stage, her three words very rapidly became many structured sentences. Seen this way, language acquisition was quite miraculous: children went from one word to many in the space of two years.

Experts now agree the picture is more complicated. Strictly speaking, there is no one-word stage. The first sign of language is usually a gesture, which infants will make at about ten months. The best way to think about this process is that it begins with a one-element stage, and that element may be a word or a gesture, such as pointing. If you have ever seen a baby sit and whack his high chair table imperiously, demanding his lunch, you have witnessed the origins of language in the individual. Following the first one-element stage, there is a two-element stage, when word and gesture appear together. This combination can function like a sentence, as when a child says "eat" and points at a banana at the same time. Gesture-and-speech combinations increase between fourteen and twenty-two months.

Children also show a three-element stage using both gesture and speech before producing three-elements in speech alone.[10] Following this stage, speech starts to emerge as the prime method of communication.

These findings suggest that gesture doesn't simply precede language but is fundamentally tied to it.[11] In fact gesture and speech are so integral to each other in children that researchers are able to predict a child's language ability at three years of age based on its gesturing at one year. They can also diagnose delays or problems that children might be having with language by examining their gestures.

For a long time the trend was to regard infants, much like animals, as mute and unthinking. Until they learned their first few words, it was thought that not a lot was going on inside their heads. And certainly, if you removed gesture from the language acquisition picture, children did seem eventually to pull language out of thin air. But when you take gesture into account, you can see the preliminary scaffolding of language even before a child has spoken a word, and the acquisition of language, while still incredible, looks a little less mysterious.

Developmental psychologists now talk about the cross-modality of language, meaning that language is expressed in various ways. Instead of the image of a brain issuing language to a mouth, from which it emerges as imperfect speech, think, rather, of language emerging in the child as an expression of its entire body, articulating both limbs and mouth at the same time.

Before the teaching of sign language became widespread, and more recently the use of cochlear implants, the fate of deaf children was contingent on their family situation. Most children who are born without hearing now receive systematic education in schools designed to help them, but there are still rare cases where children who are born deaf do not receive sign language instruction. Whether the reasons are socioeconomic or otherwise, these children are generally spoken to by their parents using normal language and gesture, and they must invent their own ways to express what they want. Susan Goldin-Meadow, who investigates gesture at her laboratory in Chicago, has studied a number of these children. The gestural language they invent is called homesign. Goldin-Meadow's work on homesign and other gestures reveals a great deal about the way the ancient platform of gesture works in modern humans.

The versions of homesign used by each of these children share a number of traits, including the fact that they generally feature a stable list of words and a kind of syntax. Certain words will appear in a particular spot in a sentence depending on the role they take. There is structure in homesign words, as well as in homesign sentences. The symbols that homesigning children invent are not specific to a particular situation or time. For example, they might use a "twist" gesture to ask someone to open a jar, or to indicate that a jar has been twisted open, or to observe that it is possible to twist a jar open. Homesign symbols are also like words in that the number that can be invented appears to be limitless, as well as stable.[12] Even though these children are exposed to a normal combination of gesture and speech by their parents, their own homesign doesn't resemble their parents' gesturing. Children who develop homesign pass through stages of development similar to those of hearing children who are learning speech. Moreover, the linear ordering of elements in a homesign utterance appears to be universal, regardless of the language community the children are born into. Interestingly, if hearing people gesture without speaking, their gestures start to look like the signs of homesigners.

How is it possible that these homesign children who are spoken to (even if they can't hear the words) and gestured at end up gesturing communicatively in the absence of a sign education? Where does this facility for structure and words come from? Goldin-Meadow believes that sentence- and word-level structure are inherent.

Altogether, Goldin-Meadow's studies show that gesture is highly versatile. It is used both with speech and without, and it differs depending on whether it is used with the spoken word. It takes a backseat when it accompanies language, and it becomes much more mimetic when it is used alone. When gesture carries the full burden of communication, says Goldin-Meadow, it becomes much more segmented. She likens it to beads on a string.

Homesign may represent an extreme example of the way that gesture and speech interact, but other recent experiments have demonstrated how speech and gesture can depend on each other. It's been shown that adults will gesture differently depending on the language they are speaking and the way that their language encodes specific concepts, like action. For example, experimenters have compared the idiosyncratic way that Turkish and English speakers describe a cartoon that depicts a character rolling down a hill. Asli Özyürek, a research associate at the Max Planck Institute for Psycholinguistics, compared the performance of children and adults in this task. She showed that initially children produce the same kinds of gestures regardless of the language they are speaking. It takes a while for gesture to take on the characteristic forms of a specific language. When it does, people change their gestures depending on the syntax of the language they are speaking. At this stage, instead of gesture's providing occasional, supplementary meaning to speech without being connected to it in any real way, language and gesture appear to interact online in expression.

In another experiment Goldin-Meadow asked children and adults to solve a particular type of math problem.[13] After they completed the task, the participants were asked to remember a list of words (for the children) and letters (for the adults). Subjects were then asked to explain at a blackboard how they had solved the problem. Goldin-Meadow and her colleagues found that when the experimental subjects gestured during their explanation, they later remembered more from the word list than when they did not gesture. She noted that while people tend to think of gesturing as reflecting an individual's mental state, it appears that gesture contributes to shaping that state. In the case of her subjects, their gesturing somehow lightened the mental load, allowing them to devote more resources to memory.

Gesture interacts with thought and language in other complicated ways. In another experiment Goldin-Meadow asked a group of children to solve a different kind of problem.[14] She then videotaped them describing the solution and noted the way they gestured as they answered. In one case, the children were asked if the amount of water in two identical glasses was the same. (It was.) One of the glasses was then poured into a low and wide dish. The children were asked again if the amount of water was the same. They said it wasn't. They justified their response by describing the height of the water, explaining it's different because this one is taller than that one. As they spoke, some of the children produced what

Goldin-Meadow calls a gesture-speech match; that is, they said the amounts of water in the glass and the bowl were unequal, and as they did, they indicated the different heights of the water with their gesture (one hand at one height, the other hand at the other height). Other children who got the problem wrong showed an interesting mismatch between their gesture and their speech. Although these children also said that the amount of the water was different because the height was different, gesturally they indicated the width of the dishes. "This information," said Goldin-Meadow, "when integrated with the information in speech, hints at the correct answer—the water may be higher but it's also skinnier."

The mismatch children suggested by their hand movements that they knew unconsciously what the correct response was. And it turned out that when these children were taught what the relationship between the two amounts of water was after the initial experiment, they were much closer to comprehension than those whose verbal and gestural answers matched—and were wrong.

Gestures also affect listeners. In another experiment children were shown a picture of a character and later asked what he had been wearing. As the researcher posed the question, she made a hat gesture above her head. The children said that the character was wearing a hat even though he wasn't.

Such complicated dependencies and interactions demonstrate that speech and gesture are part of the same system, say Goldin-Meadow and other specialists. Moreover, this system, made up of the two semi-independent subsystems of speech and gesture, is also closely connected to systems of thought. Perhaps we should designate another word entirely for intentional communication that includes gesture and speech. Whatever it should be, Goldin-Meadow and others have demonstrated that this communication is fundamentally embodied.[15]

The most important effect of this research is that it makes it impossible to engage with the evolution of modern language without also considering the evolution of human gesture. Precisely how gesture and speech may have interacted since we split from our common ancestors with chimpanzees is still debated. Michael Corballis, who wrote *From Hand to Mouth: The Origins of Language,* has suggested that quite complicated manual, and

possibly facial, gesture may have preceded speech by a significant margin,
arising two million years ago when the brains of our ancestors underwent
a dramatic burst in size. The transition to independent speech from this
gesture language would have occurred gradually as a result of its many
benefits, such as communication over long distances and the ability to use
hands for other tasks, before the final shift to autonomous spoken language. Other researchers stress how integral gesture is to speech today, arguing that even as the balance of speech and gesture may have shifted
within human communication, it is unlikely that gesture would have
evolved first without any form of speech. David McNeill, head of the well-
known McNeill Laboratory Center for Gesture and Speech Research at
the University of Chicago, and colleagues propose that from the very beginning it is the combination of speech and gestures that were selected in
evolution. What about the other side of the coin—what about speech? It is
not as ancient as gesture, but when did it evolve? And how closely related
is speech to the vocal communication of other animals?

8. You have speech

E ven though more research has been conducted on primate vocalizations than on primate gesture, it has been considerably less productive.
Vocalization in nonhuman animals is much less flexible than gesture. Most
vocalizations, like alarm calls, seem to be instinctive and specific to the
species that produces them. Many kinds of animals that are raised in isolation or fostered by another species still grow up to produce the calls of
their own kind. Researchers at the Neurosciences Institute in San Diego
transplanted brain tissue from the Japanese quail to the domestic chicken;
the resulting birds, called chimeras, spontaneously produced some quail
calls as they matured.[1] And unlike human talkers, vocalizing animals seem
to be pretty indifferent to their listeners. Vervets, for example, typically
produce alarm calls whether there are other monkeys around or not. Even
though we still have a lot to learn about calls in the wild, it appears that
there are relatively few novel calls in ape species. What's more, apes don't
seem to make individually distinctive calls, even though other monkeys—
which are more distantly related to us—do.

One of the biggest differences between ape gesture and vocalizing is
that many communicative gestures appear to be voluntary and intentional
in a way that sound is not. Still, the involuntary nature of animal vocalizations has been somewhat exaggerated. It is said, for example, that when
apes make a sound it is always an emotional response and not really generated by choice (in contrast with gesture, which is demonstrably voluntary). In recent years, this position has had to shift to accommodate some
interesting findings about the rudiments of control in the vocal domain.
Evidence exists, for example, that chimps can suppress calls in dangerous
situations where a loud noise would draw attention to them. Some orangutans make kissing sounds when they bed for the night. Kissing is not

instinctive, it's volitional—one of those cultural traditions that distinguish groups within a species from one another.

In a recent experiment Katie Slocombe and Klaus Zuberbühler (who earlier demonstrated the ability of zoo chimpanzees to distinguish between types of food with wordlike calls) found that wild chimpanzees seem to adjust their screams based on the role they play in a fight. The researchers looked at two different types of screams in the wild chimpanzees of the Budongo Forest in Uganda. In a conflict situation, the animals typically produce a victim scream, in which the pitch is very consistent, and an aggressor scream, where the pitch varies, with a fall at the end. Other chimps appear to use this information, said Slocombe. The researcher witnessed one exchange in which a young male was harassing a female chimp that was giving loud victim screams in response. At one point, said Slocombe, the female had clearly had enough and began instead to make aggressor screams back at the young male. She was then joined by another female in retaliating against the male. The second female appeared from out of sight, so she must have used the information in the first female's scream to make her decision. "Normally," said Slocombe, "chimpanzees will see parts of the fight, and therefore it is impossible to tell if they are attending to the information in the screams or just what they see."

Slocombe was interested in establishing whether any particular information about a given situation was reliably communicated by the chimpanzee screams. She recorded examples of victim screams and noted the circumstances in which they occurred. An analysis of her recordings showed that it was possible to distinguish from the screams alone between high-risk situations and low-risk ones. In the first case, the screams tended to be long and high-pitched, whereas in low-risk situations the screams were shorter and lower in pitch.

There are other intriguing connections between the way we use our mouths and the way other apes do. Researchers have noted a peculiar feature of gesture that appears to be shared between humans and chimpanzees. Imagine a child learning how to write, his hand determinedly grasping the pencil and his tongue sticking out of the side of his mouth. Or visualize a seamstress biting her lips as she sews a small thread. Such unconscious mouth movements often accompany fine hand movement in humans. Of course, mouth and hand movements co-occur with speech and gesture, but in this case it seems that the mouth movement follows the hands (not the other way around). Experiments have shown that fine motor manipulation of objects by chimps is often accompanied by sympathetic mouth movements. The finer the hand movements are, the more chimps seem to move their mouths. David Leavens suggests that the basic connection between mouth and hand in primates could date back at least fourteen million years, to the common ancestor of human and orangutan.

Despite such new insights into the utterances of other apes, a vast gap remains between the apparent vocal abilities of all primates and the speech abilities of human beings. Speech starts simply enough with air in the lungs. The air is forcefully expelled in an exhalation, and it makes sound because of the parts of the body it blows over and through—the vibrating vocal cords, the flapping tongue, and the throat and mouth, which rapidly opens and closes in an odd, yapping munch. It's easy to underestimate the athletic precision employed by the many muscles of the face, tongue, and throat in orchestrating speech. When you talk, your face has more moves than LeBron James.

It takes at least ten years for a child to learn to coordinate lips, tongue, mouth, and breath with the exacting fine motor control that adults use when they talk. To get an idea of the continuous and complicated changes your vocal tract goes through in the creation of speech, read the next paragraph silently, letting your mouth move but making no sound—just *feel* the process.

What's amazing about speech is that when you're on the receiving end, listening to the noise that comes out of people's mouths, you instantaneously *hear* meaningful language. Yet speech is just sound, a semicontinuous buzz that fluctuates rapidly and regularly. Frequencies rise and fall, harmonics within the frequencies change their relationships to one another, air turbulence increases and dies away. It gets loud, and then it gets quiet.

The rate of the vocal cords' vibration is called the fundamental frequency, an important component of speech. Perhaps the most significant aspects of the sound we make are the formant frequencies, the set of frequencies created by the entire shape of the vocal tract. When you whisper, your vocal cords don't vibrate and there is no fundamental frequency, but people can still understand you because of the formant frequencies in the sound.

Overall, the variations in loudness, pitch, and length in speech that we think of as the intonation of an utterance help structure the speech signal while also contributing to its meaning. Prosody, the rise and fall of pitch and loudness, can be emotional, it can signal contrast, and can help distinguish objects in time and space ("I meant *this* one, not *that* one"). Prosodic meaning can be holistic, like gesture. It can signal to the listener what a speaker thinks about what he is saying—how sure he is of something, whether it makes him them sad or happy. When people make errors in what they are saying, they can use intonation to guide listeners to the right interpretation. Prosody can also mark structural boundaries in speech. At the end of a clause or phrase, speakers will typically lengthen the final stressed syllable, insert a pause, or produce a particular pitch movement.

Even though we hear one discrete word after another when listening to a speaker, there's no real silence between the words in any given utterance, so comprehension needs to happen quickly. Whatever silence does fall between words does so mostly as a matter of coincidence—as a rule, when sounds like *k* and *p* are made (like at the beginning and end of "cup"). These consonants are uttered by completely, if briefly, blocking the air flowing from your lungs. (Make a *k* sound, but don't release it, and then try to breathe.) So while a sentence like "Do you want a cup of decaffeinated coffee?" may be written with lots of white space to signify word breaks, the small silences within the sound stream don't necessarily correspond to the points in between words.

The beginning of speech is found in the babbling of babies. At about five months children start to make their first speech sounds. Researchers say that when babies babble, they produce all the possible sounds of all human languages, randomly generating phonemes from Japanese to English to Swahili. As children learn the language of their parents, they narrow their sound repertoire to fit the model to which they are exposed. They begin to produce not just the sound of their native language but also its classic intonation patterns. Children lose their polymath talents so effectively that they ultimately become unable to produce some language sounds. (Think about the difficulty Japanese speakers have pronouncing English *l* and *r*.)

While very few studies have been conducted on babbling in humans, SETI (Search for Extraterrestrial Intelligence) Institute researcher Laurance Doyle and biologist Brenda McCowan and colleagues discovered that dolphin infants also pass through a babbling phase. (In 2006 German researchers announced that baby bats babble as well.) In the dolphin investigation Doyle and McCowan used two mathematical tools known as Zipf's law and entropy. Zipf's law was first developed by the linguist George Zipf in the 1940s. Zipf got his graduate students to count how often particular letters appeared in different texts, like *Ulysses,* and plotted the frequency of each letter in descending order on a log scale. He found that the slope he had plotted had a –1 gradient. He went on to discover that most human languages, whether written or spoken, had approximately the same slope of –1. Zipf also established that completely disordered sets of symbols produce a slope of 0. This meant there was no complexity in that particular text because all elements occurred more or less equally. Zipf applied the tool to babies' babbling, and the resulting slope was closer to the horizontal, as it should be if infants run randomly through a large set of sounds in which there is little, if any, structure.

When Doyle and McCowan applied Zipf's law to dolphin communication, they discovered that, like human language, it had a slope of –1. A dolphin's signal was not a random collection of different sounds, but instead had structure and complexity. (Doyle and his colleagues also applied Zipf's law to the signals produced by squirrel monkeys, whose slope was not as steep as the one for humans and dolphins (–0.6), suggesting they have a less complex form of vocalization.[2] Moreover, the slope of baby dolphins' vocalizations looked exactly like that of babbling infants, suggesting that the dolphins were practicing the sounds of their species, much as humans do, before they began to structure them in ordered ways.

The scientists also measured the entropy of dolphin communication. The application of entropy to information was developed by Claude Shannon, who used it to determine the effectiveness of phone signals, by calculating how much information was actually passing through a given phone wire. Entropy can be measured regardless of what is being communicated because instead of gauging meaning, it computes the information content of a signal. The more complex a signal is, the more information it can carry. Entropy can indicate the complexity of a signal like speech or whistling, even if the person measuring the signal doesn't know what it means. In fact, SETI plans to use entropy to evaluate signals from outer space: if we ever receive an intergalactic message and can't decode its meaning, we can apply entropy to give us an idea about the intelligence of the beings that transmitted the signal even if we can't decode the message itself.

The entropy level indicates the complexity of a signal, or how much information it might hold, such as the frequency of elements within the signal and the ability to make a prediction about what will come next in the signal, based on what has come before. Human languages are approximately ninth-order entropy, which means that if you had a nine-word (or shorter) sequence from, say, English, you would have a chance of guessing what might come next. If the sequence is ten words or more, you'll have no chance of guessing the next word correctly. The simplest forms of communication have first-order entropy.[3] Squirrel monkeys have second- or third-order, and dolphins measure higher, around fourth-order. They may be even higher, but to establish that, we would need more data. Doyle plans to record a number of additional species, including various birds and humpback whales.

Many of the researchers interviewed for this book would stop in the middle of a conversation to illustrate a point, whether it concerned the music of protolanguage or the way that whales have a kind of syntax, by imitating the precise sound they were discussing. Tecumseh Fitch sat at a restaurant table making singsong *da-da da-da da-DA* sounds. Katy Payne, the elephant researcher, whined, keened, and grunted like a humpback whale in a small office at Cornell. Michael Arbib, the neuroscientist, stopped to purse his lips and make sucking sounds. In a memorable radio interview, listeners heard the diminutive Jane Goodall hoot like a chimpanzee.

As well as demonstrating the point at hand, the researchers' performances illustrated on another level one of the fundamental platforms of language—vocal imitation. Imitation is as crucial to the acquisition of speech as it is to learning gesture (another way in which these systems look like flip sides of the same coin). Humans are among the best vocal imitators in the animal world, and this is one area in which we are unique in our genetic neck of the woods. Even though chimpanzees do a great job of passing on gestural traditions and tool use in their various groups, they don't appear to engage in a lot of imitation of one another's cries and screeches. Orangutans must have some degree of imitation in the vocal domain, otherwise they couldn't have developed the "goodnight kiss" tradition. But humans have taken the rudiments of this ability and become virtuosos.

It would appear that this skill has become fully developed in our species over the last six million years, since we split from our common ancestor with chimpanzees and bonobos. From the babbling stage on we start to repeat simple vowels and consonants, like "mamamamamama," advancing to whole words, sentences, longer tracts, all the while using rhythms, pitch, and loudness. Still, like many of our other abilities, this one is built on a platform that stretches back a long way in evolutionary time.

Vocal learning is one of the reasons that Fitch believes the field of language evolution is worth pursuing. "Where you get any kind of open-ended learning, you have the ability to pair signals with meaning. And we didn't have to evolve that, because our common ancestor with other primates already evolved it. What we don't have in a chimp or any other ape is vocal learning—the ability to generate new signals. Dogs, for example, aren't able to invent new barks."

Some other animals are also exceptional at vocal imitation, whether it involves imitating a human or a member of their own species. Songbirds are not born with genetic programs from which their songs arise. Instead, in the same way that we are born with a predisposition to produce the sounds of language, the specifics of which we still must learn, they need to be exposed to the songs of their species in order to acquire them.[4]

African gray parrots, Alex's species, as well as other types of parrots, are well known for their excellence in imitating human words. Some animals seem to entertain themselves by imitating the sounds of inanimate objects. Mockingbirds have been heard imitating sounds like car alarms and mobile phones, and elephants in Kenya have been recorded making almost perfect reproductions of the sound of trucks from a road nearby. Whales are very good at vocal learning. Each mating season, the males

come together to sing, riffing on the songs of the previous season and producing something new from them. Dolphins are as talented at vocal imitation as they are at gestural imitation. As Lori Marino explained, "They
seem to be able to imitate a number of different dimensions of a behavior.
They can imitate the physical dimension, but also the temporal dimension.
They can imitate rhythms. For instance, you can give them a series of
tones, and they'll be able to imitate the rhythm of that series of tones. So
if you give them ENH-ENH, ENH-ENH-ENH, ENH-ENH, they'll give
you ENH-ENH, ENH-ENH-ENH, ENH-ENH."

There have been odd, one-off cases of individual animals showing exceptional imitative talents. Fitch is fascinated by the story of Hoover, a harbor seal at the New England Aquarium that was raised by a Maine
fisherman. Hoover surprised visitors by saying, "Hey, hey, you, get outta
there!" Hoover didn't "talk" until he reached sexual maturity, but once he
started, he improved over the years. He spoke only at certain times of the
year (not as much in the mating season) and would reputedly adopt a
strange position in order to do so. He didn't move his mouth. In *The Symbolic Species,* Terrence Deacon recounts stumbling across Hoover while
walking near the aquarium one evening. He thought a guard was yelling at
him ("Hey! Hey! Get outta there!"). Deacon reports that Hoover died unexpectedly of an infection and his body was disposed of before his brain
could be examined.

"We don't know if Hoover was a mutant or if other seals can do this,"
said Fitch. "It's not hard to train a seal to bark on command. There's a sea
lion named Guthrie at the New England Aquarium. He gets rewarded
when he does something different. His barks are not very special, but they
are bona fide novel vocalizations." Fitch relates Hoover's ability to the
Celtic selkie myths, which may have originated in earlier Hoover-like accounts. "It's not uncommon for humans to take seals into their homes,"
said Fitch. "Maybe we just need to expose male seals to human speech and
the right social context," and they'll be able to learn some speech.

What makes Hoover so interesting, according to Fitch, is that all the
other animals that are excellent at vocal learning, with the possible exception of bats, use a completely different process from the ancestral vertebrate mechanism for making sound. What we use for vocal production is

the same thing that a frog uses—a larynx and tongue, equipment that has been around since early vertebrates dragged themselves onto land. Birds, on the other hand, have evolved a completely novel organ—the syrinx. The toothed whales, like dolphins and killer whales, have evolved a unique organ in their nose, and we still don't really know how other whales make sound. "It's hard to peer down the nostril of a humpback or get them in an X-ray setup while they are singing," observes Fitch.

Early speech researchers like Philip Lieberman proposed that one of the adaptations that humans made to produce language and speech was a descended larynx. The human larynx is a complicated assemblage of four different kinds of cartilage and the small, bent hyoid bone that sits upon them. The area above the larynx is called the upper respiratory tract. Below the larynx there are two tracts: the windpipe, which leads to the lungs, and the digestive tract, leading to the stomach. When humans swallow, the larynx essentially closes, ensuring that food or liquid doesn't fall into our lungs. The larynx also contains the vibrating vocal cords we use in speech.

In many animals, such as other apes, the larynx sits high in the throat. In fact, for most animals the larynx is positioned so high that it's effectively in the nasal passages, meaning that these creatures can breathe and drink at the same time. Human babies, who are born with high larynxes, can do the same, but by the time they turn three, the larynx has descended and this is no longer possible. For boys, the larynx descends a bit more in adolescence, giving their voices a more baritone timbre. Somewhere in our evolutionary history—between the present and the last common ancestor we had with chimpanzees and bonobos six million years ago—our larynx dropped, making the upper and lower respiratory tracts roughly equal in size. It is these two tubes that allow humans to make such a wide range of different vowel and consonant sounds.

For a long time researchers thought that the descended human larynx was the smoking gun of speech evolution, but the picture turns out to be more complicated than that. Most previous findings about the larynx of other animals were based on the anatomy of dead specimens, but Fitch investigated the behavior of living, vocalizing animals and discovered that the larynx is a far more mobile structure than previously thought. He found that other animals that don't have a permanently descended larynx pull it into a lower position when they vocalize. Dogs do so, as do goats, pigs, and monkeys. In addition, Fitch discovered that some animals have a permanently descended larynx, including species as diverse as the lion and the koala. What this means, said Fitch, is that you can't assume that the reason the larynx descended in humans was for speech; you have to be able to explain the function of the descended larynx in these other animals as well.

In his Ph.D. work Fitch demonstrated a basic correlation between body size and the deepness of voice. In the animal kingdom this correlation provides extremely useful information. If you hear a competitor wooing the female you are interested in, and you can tell from his voice alone that he is much bigger than you, slinking away without direct confrontation makes the most evolutionary sense. Fitch argues that this is how we initially came by our descended larynx, meaning that one of the fundamental elements of our ability to create speech came about not because of language but as a primitive mechanism to signal an exaggerated body size.

Other critics maintain that the descended larynx is most likely an example of evolutionary adaptation in the human lineage. Steven Pinker explained:

> I think it's premature to say that there has been no evolutionary change in speech perception and speech production mechanisms. In fact, certainly for speech production mechanisms I think the argument that there's been no adaptation or evolutionary change is very weak. It's based on the idea of the descent of the larynx seen in some other mammals, which did not evolve it for language, but rather for bellowing in a more macho way. So yes, it's marginally possible that the larynx descended in humans for some reason other than language, but that theory doesn't work for humans, because we have a descended larynx in both sexes, where exaggerating body size by bellowing more loudly is not a factor.

Fitch adds that just because the descended larynx may have come about for reasons other than speech doesn't mean it wasn't then co-opted—or in

Darwinian terms, exapted—for speech evolution. He emphasizes the possibility of gradual evolution. "The fact remains," he writes, "that the human larynx is unusual (though not unique) among mammals." It's possible, he says, that early hominids had a mobile larynx, like those of dogs and pigs. But as they began to develop the extensive sound range of speech, it became more efficient to leave the larynx in the descended position instead of pulling it back to vocalize, as other animals do.[5]

The notion of a graded evolutionary descent is supported by recent findings on the larynx of chimpanzee infants, which also undergoes a process of descent. This process results from a somewhat different mechanism, accomplished by the descent of the skeleton around the chimpanzee hyoid bone rather than the descent of the hyoid bone itself. Nevertheless, it suggests that descent of the larynx in humans is unlikely to have occurred in one big, speech-related transition.[6]

Other features of vocal production in humans that appear to be especially attuned for language include a particular kind of muscle fiber in the vocal folds. According to Ira Sanders at the Mount Sinai School of Medicine, slow tonic muscle fibers have unique features. They don't twitch like most muscle fibers but contract in a precise, graded fashion. Sanders examined a series of adult tongues and found that the slow tonic muscle fibers occur there in high numbers. Other mammals do not have this kind of muscle in their vocal folds.

Attempts to find fossil evidence for the key anatomical changes required for modern human speech have been mostly unsuccessful. Fitch attributes this to the fact that "the vocal tract is a mobile structure that essentially floats in the throat, suspended from the skull by elastic ligaments and muscles." Some researchers have compared the part of the spine that affects voluntary breathing—a crucial part of speech production—in *Homo sapiens, Homo ergaster,* and earlier hominids. It appears that this region is significantly enlarged in modern humans as compared with earlier ancestors.[7]

Regardless of their other theoretical differences, most language evolution researchers agree that human speech appears to have evolved in the last six million years to meet some of our species' unique communication needs. The most basic and obvious evidence for this is that despite concerted efforts to teach spoken language to other primates, no attempt has been successful. At most, chimpanzees have been trained to utter a few words.[8] But the perception of speech is another matter.

The human facility for perceiving speech begins very young: small babies have been shown to prefer the sounds of speech to nonspeech sounds. It is a fascinating paradox that humans can hear only up to fifteen different non-speech sounds per second, and beyond this they hear unremitting noise. Yet when they decode speech, they hear twenty to thirty distinct sounds per second. Somehow human speakers can pack, and in turn unpack, almost twice as many sounds if those sounds consist of consonants and vowels that are the components of the language they speak.

Humans also have a remarkable ability to calibrate the way that speak-ers' voices occupy many different spots within the range of possible pitch. Children's voices are typically the highest, women's are in the middle of the range, and men can have very deep timbre.[9] This means that even though they are all speaking the same language, the formant frequencies of any given vowel can be quite different. Nevertheless, we understand the speakers of our language to be making the same sounds.

Some researchers believe that the movements of our throats, tongues, mouths, and faces in speech are as important as the sound of speech. They hold that at some level, speech is also gesture. Indeed, our ability to per-ceive the speech of others is based in part on our knowledge of the motor movements we make when we produce it. It's been demonstrated that sub-jects who are shown a video of someone saying "ga" that is accompanied by a recording of the sound "ba" perceive something entirely different. They will "hear" "da," which in terms of speech production is in between the "ga" and "ba" sounds ("ba" is made with the lips, "da" is made with the tongue touching the roof of the mouth behind the teeth, and "ga" is made with the back of the tongue hitting the roof at the back of the mouth). This phenomenon is called the McGurk effect, and it demonstrates that as far as the perception of such simple sounds goes, people can be as influenced by the motor acts they see as by the sound they hear.

One of the most important strategies that human brains use to under-stand speech is called categorical perception. Even though we think of the sounds in our alphabet as being distinct from one another, there is a continuum between sounds like *p* and *b,* which differ only in the timing of the vocal cords' vibrations.

Scientists who first discovered categorical perception in the 1950s found that timing is critical in the perception of sound. For example, listeners' perception of *b-p* changes at the twenty-five-millisecond mark. If they hear the *b-p* sound and the vocal cords begin to vibrate at 10 or 20 ms, they hear a *b;* if the vocal cords begin to vibrate at 25 ms or higher, even though everything else about the sound is the same, they hear a *p* instead. It is as if a switch is thrown at the 25 ms mark. People hear only one sound or the other, not a sound that is a little like both. In the 1970s the experiment was repeated using infants as subjects, and researchers found that children make the same categorical distinction between sounds. The finding was hailed as evidence of an innate and uniquely human language trait.

That claim was made without any relevant data from animal studies, and it took only a few years to be invalidated. In 1975 two researchers repeated the infant study but used chinchillas, which also proved to have categorical perception. So even though this trait fundamentally underlies the human ability to perceive speech, it's a much more general feature of animal auditory systems. Later experiments have shown that categorical perception also applies to nonspeech sounds.

Other important properties of human speech perception are shared by other animals. In a study conducted by Marc Hauser and colleagues, researchers found that humans aren't the only species with the ability to identify different languages based on their characteristic rhythms. Tamarins, tiny primates that roam the forests of the Amazon basin, can distinguish between languages based on different rhythmic cues.[10] This ability suggests that we probably didn't evolve our sensitivity to linguistic rhythm for the specific purpose of understanding or producing speech, even though that is now its primary function. Instead we use a general perceptual mechanism that is shared among animals. In another study Hauser and colleagues extended the earlier findings to show that other properties of this perceptual mechanism are common to humans and tamarins. For example, neither human babies nor tamarins distinguish between languages that come from the same rhythmic class, such as English and German, or that are rhythmically similar like English and Dutch. However, they could tell the difference between rhythmically different languages like Japanese and Polish. Another property of speech perception is the ability to hear the formant frequencies that characterize different vowels. In another study, Hauser and colleagues have pointed out that some animals are able to use formant frequencies to make distinctions between sounds and that other species perceive formants in their own species' vocalizations.[11]

Many questions remain about the animal perception of speech. There is no evidence that animals either have or could be trained to develop the ability to parse out the vast number of words in the semicontinuous speech stream of human conversation. Still, we have yet to explain the very basic fact that animals like the Border collie Rico, the African gray parrot Alex, and the bonobo Kanzi clearly have some capacity for perceiving and understanding words within a semicontinuous speech stream. These animals appear to take the speech-noise, identify distinct sounds within it, break the whole thing up into smaller meaningful units (if not as many as humans, then at least some), and derive a meaning from that. Kanzi, for example, has learned that the buzz coming out of someone's mouth can be broken up into recognizable units ("throw," "ball," "water") that can be combined to create larger meaningful units ("Throw the ball in the water").

In order to accurately determine how much of speech perception is shared by humans and animals, researchers must eventually explain how these creatures adjust to different speakers in the same way that humans do and, even if one person's p is different from another's, still make sense of the word, no matter who is saying it.

Of course, humans do a lot more perceptually than simply pulling a few words out of a larger set of vocalizations. We parse the speech stream exhaustively, and we do it in real time, picking out sounds that are jammed many to a second. We identify the words they create and at the same time the sentences they create. "Speech flows together like this" actually sounds more like "Speechflowstogetherlikethis," and yet we effortlessly work out where one word has ended and another has begun in real time.

Researchers like Marc Hauser and Tecumseh Fitch believe that the claims for human uniqueness have been proven wrong so often in the perceptual domain that people should no longer make default assumptions about any special human ability. In their view, it is reasonable to believe that the hearing part of language is completely shared with many other animals. But others are more skeptical.

Speech perception is such a complicated task, Steven Pinker pointed out, that even speech recognition systems on today's modern computers require that you talk to them with exaggerated breaks between words unless they are trained on a specific person's voice. "Understanding connected speech from a variety of speakers is a remarkable ability," he said, "one that artificial intelligence researchers have had enormous difficulty duplicating in computers. It certainly has not been shown that other animals are capable of processing continuous speech. It would be very hard to test, because they don't have the language that continuous speech is converted into. The fact is that we don't know that they can do it, and I'd be very skeptical if they can."

9. You have structure

Although many components of language have some kind of analog in animal communication, our close relatives typically lack highly structured signals. Of course birdsong can be complexly patterned, but ape and monkey communication seems to consist mostly of unanalyzable cries. Human language involves two types of structures. In the first, elements from a finite set of meaningless sounds are combined into meaningful words and parts of words, known as morphemes. Linguists call this phonology. The rules of phonology cover intonation and rhythm as well as the way specific sounds can be combined. The rules of sound apply at the smallest scale, between two single sounds that occur side by side, and over vast tracts of speech—from single sentences that either rise or fall depending on whether they are questions, to lengthier statements that end on a falling intonation. All these rules change depending on the language that is spoken.

In the second type of structure, words and morphemes are combined into phrases. This is what linguists call syntax. In 1960 the linguist Charles Hockett said that the relationship between the two types of combinatory rules was one of the major design features of human language; he called it "duality of patterning."

Inevitably, both kinds of structure have been found to be not restricted to humans. Elements of phonology operate not just in birdsong but in the songs of whales. Phrases in these songs recur and are used again. In one early experiment Marc Hauser and a colleague demonstrated that vervet monkeys use a fall in pitch to mark the end of an utterance and that other vervets seem to interpret this as a signal to take a turn in vocalizing, like humans do. Tecumseh Fitch suggests there may be other elements of sound rules that animals share. Rhythm is an important element of human language, and Fitch points to the rhythm in the dominance displays of chimpanzees and gorillas as a possible precursor for this ability in humans. Gorillas put on impressive performances of vocalizing and rhythmic chest beating, and while this behavior has been little studied, it might provide a clue to the origins of rhythm in humans. Still, chimpanzees do not speak, and neither do they dance. If important analogs for this aspect of language exist in other animals, there are also important distinctions. Not only does other animal vocal communication not have the range of distinct sounds of human language, it doesn't appear to employ anything like the number and range of rules that we have for combining speech sounds.

Interestingly, it's been pointed out that the rules of phonology contradict Chomsky's notion of the poverty of stimulus—the idea that there is not enough information in the language a child hears for it to learn language. Philip Carr, a phonologist at the University of Montpellier in France, says there is abundant evidence of the rules of phonology in the speech that children hear. The "data are more than complete," he wrote. Neonates, according to Carr, have access to more information than they need to understand the sound system of their language.[1]

Of the two types of structures, syntax has been the more hotly contested in the language evolution debate. At its most basic, syntax is a series of rules for combining words in a meaningful way. All the words in the following sentence make perfect sense by themselves, but because the way they are lined up defies the syntax of English, there is no larger meaning: *the the are up way they meaning lined there no syntax English is defies larger of.* Until very recently it was believed only we could understand or deploy any of the structural devices found in human syntax, but Kanzi showed that this is not entirely the case. He is able to learn and apply some rules to structure the symbols with which he communicates. In addition, Klaus Zuberbühler has also established that rudimentary syntax can occur in the natural cries of monkeys in the wild.

Different types of syntax have been observed in the communication of a number of primate species. The black-and-white colobus, the titi monkey, the male gibbon, the chimpanzee, and the wedge-capped capuchin monkey have combinations of calls in their repertoire of cries. The black-

and-white colobus uses a snort as an alarm call, but also places it before a roar, a combination that is used to help groups of these monkeys keep their distance from one another. The titi monkey combines several different calls into various combinations, and the response of its listeners shows that they distinguish between the different ordering of the sounds. Gibbons arrange a series of sounds into structured vocalizations, and the same is true of capuchins. In the case of gibbons, when the animal's song is arranged in a normal order, the listening gibbons squeak in response.

Zuberbühler wanted to know whether an obvious change of meaning resulted from the way that elements of the calls were ordered. He started with the Campbell's monkey in the Taï Forest of the Côte d'Ivoire. Like vervets, these animals employ different kinds of alarm calls, with one distinctive cry to warn of crowned-hawk eagles and another for leopards. They also use an interesting combination cry, in which one of the alarm calls is preceded by a boom sound. Boom-plus-alarm combinations appear to indicate a lesser threat, and are used in a situation that calls for a response to the alarm cries of a distant group, the detection of a far-off predator, or less direct dangers like falling trees or breaking branches.[2]

Zuberbühler had shown in earlier experiments that Diana monkeys respond to the cries of other species. Even though the calls of the Diana monkeys are very different from those of the Campbell's monkey, the Diana monkeys, who live closely with the Campbell's monkeys, appear to both understand and use their alarm cries to protect themselves. For example, if it hears a Campbell's monkey make an alarm call for an eagle, a Diana monkey will make its own distinct eagle alarm cry. In the syntax experiment, Zuberbühler played a series of Campbell's monkey alarm calls to a group of Diana monkeys. The recordings consisted either of Campbell's monkey alarm calls or the Campbell's monkey phrase, boom-plus-alarm. (In order to run the experiment, Zuberbühler had to use a great deal of stealth, approaching the monkeys without detection; otherwise he would have just provoked a series of human-induced alarm calls.) Zuberbühler confirmed that the Diana monkeys responded to Campbell's monkey alarm cries with alarm cries of their own. If he played an eagle alarm call, they'd respond with their own eagle alarm call; if he played the leopard alarm call, they would start making leopard alarm calls themselves. If Zuberbühler played a boom and then one of the Campbell's monkey alarm cries, the Diana monkeys wouldn't respond with one of their own alarms—indicating that they understood the nondirect nature of the threat.

Zuberbühler likens the boom to qualifiers in our own language, such as "maybe" and "kind of." His study, he says, suggests that primates have some naturally occurring syntactic abilities, and also suggests that projects in which animals are trained by humans to use syntax are tapping into abilities that occur naturally in these species.

In a more recent experiment, Zuberbühler and Kate Arnold showed that male putty-nosed monkeys combine two basic calls to add meaning to a message. Typically, these monkeys produce a *pyow* sound in various situations, most often as an alarm in response to the sighting of a leopard. They also make a *hack* sound when an eagle has been seen. Zuberbühler and Arnold discovered that male putty-nosed monkeys also make a *pyow-hack* sound, a combination call that signals that either a leopard or an eagle has been seen. The difference in response is that shortly after a *pyow-hack* is made, the whole monkey troop will move location, suggesting that it has the additional message of *"Move!"*

Gibbons structure units of sound to create meaning, but their vocalizations are quite different from those of most other primates; they produce complex songs, communicating over distances up to one kilometer away. Typically gibbons form monogamous pairs, and every morning mated pairs sing a duet that pronounces their bond to neighboring apes.

Zuberbühler and colleagues recorded white-handed gibbons at Khao Yai National Park, Thailand, and found that the gibbons use their songs to repel predators as well as to perform duets. The duets and the predator songs used the same notes ("wa," "hoo," "leaning wa," "oo," "sharp wow," "waoo," and "other"), but they systematically differed in how they were arranged. At the beginning of a song, there were fewer "leaning wa" notes and significantly more "hoo" notes if a predator had been sighted. In addition, predator songs had more "sharp wows" in them and were longer overall than duets. Male- and female-specific parts of the songs also differed depending on the referent. While female-specific parts came later in the predator song, the males replied earlier to females in these songs than in the duets. As with the other Zuberbühler experiments, it was also found that the structured utterances were meaningful to neighboring animals. Nearby gibbons responded differently to the two kinds of songs.

The scientists don't view the gibbon songs as sentences created with syntactic rules about word order. There is no context in which to determine whether notes have smaller discrete meanings, like words, which build a larger meaning when combined in different ways. What is important about the gibbon utterances is that they use combinatorial rules to functionally refer to different things. The same set of sounds has two different meanings when ordered in different ways.

The simple structural rules that these primates use in the wild contradict the idea that creating meaning with structure is a special human ability. Though there remains a wide gulf between what we do with structure and what other animals do, at least some elements of our ability seem to be graded. Robert Seyfarth and Dorothy Cheney, the researchers who pioneered the vervet monkey work, suggest that more evidence for an evolutionary precursor to human syntax may be found somewhere other than in the vocal domain.

After their vervet work, Seyfarth and Cheney began to study a baboon group in the Okavango Delta of Botswana. Baboons—Old World monkeys—typically live in stable groups of 50 to 150 animals. They have a small and limited set of calls, which are largely innate, and they have no call combinations.[3] There are 80 to 90 baboons in the Seyfarth-Cheney group, and every day since 1992 someone has observed the animals. By now, Seyfarth, Cheney, and their colleagues recognize all the animals individually. The rules of baboon society, said Seyfarth, are similar to those of Jane Austen's: be nice to your relatives, and get in with the high-ranking family. For the researchers this extended period of observation has been like watching a long-running soap opera.

Baboons have a matrilineal society. Females stay in the group into which they are born, while somewhere between the ages of six and nine the males leave and join another group. Each baboon family is ranked from highest to lowest in the troop, and within each of those families each baboon is also ranked for dominance. What this means practically, said Seyfarth, is that within each group, there is one baboon that can go wherever she wants, eat whatever she wants, and sit wherever she pleases. All the other baboons will give way to her. Then there is a second baboon that can do all the same things, except with respect to the top baboon, and so it goes down the dominance hierarchy to the lowest-ranked baboon in the group. Seyfarth and Cheney have found over the years that the dominance ranks within families are as stable as those between families. Some families will, en masse, give way to other families, while within families there is a number one baboon, a number two baboon, and on, until the last baboon.

Some vocalizations are universal within the baboon group. For example, all the baboons seem to grunt all the time. Also, some cries are given in only one direction—up or down the dominance ranking. Screams and fear barks are given only to those higher in rank, and threat grunts only to those lower in rank than the grunter. The scientists also found that baboon calls are individually distinctive. Because of this, a third-party baboon can tell a great deal about the social dynamic in a group of animals just by listening to an exchange between them—he can tell which is more dominant and which individuals are involved, and therefore to what family they belong.

The researchers and some of their colleagues decided to exploit this information in an experiment in which they recorded baboon utterances and then played them back to baboon listeners. For example, they played the threat grunts and fear barks of two baboons that would normally give these kinds of calls to each other (the threat grunter was higher in dominance than the barker), and they also manipulated the recordings so that an interaction sounded as if it defied social order—a lower-ranked baboon threat-grunted at a higher-ranked baboon, and it fear-barked back.

Experiments like this have found that when played a "normal" interaction, the baboon listeners will either ignore it or look at the source of the sound for a short amount of time: the dynamic is normal to them and doesn't arouse surprise or require further investigation. When a baboon looks longer, it suggests that what it just heard has caught its attention and violated its expectations in some way, as in the case of the vocalizations that subverted the baboons' ranking. The baboon listeners looked longer for the source when the interaction violated normal expectations. These results confirmed that individual baboons recognized the ranking of others.

Seyfarth and Cheney's team also wanted to know if individual baboons understood the family rankings in the group. Two researchers waited until two baboons of different families were sitting near each other. First they played them a recording of a high-ranking baboon arguing with a low-ranking baboon, both from different families than those of the observers. Typically enough, the sitting baboons paid little attention. Then, a few weeks later, the experimenters played a recording of a fight between an unrelated baboon and a baboon from the same family as the high-ranking listener. In this case, the low-ranking listener looked up at the higher-ranked baboon, as if to see what she would do next. The researchers later played a recording of a fight between a family member of the high-ranking baboon and a family member of the low-ranking baboon. Immediately, the listeners looked at each other, indicating their awareness of the family relationships.

The researchers took the experiment further. Did the baboons memorize the dominance ranking of all ninety members of their troop? Or was it possible they were factoring in family relationships as well? Were they using some kind of mental shortcut to collapse across the ranked list of almost one hundred individuals, just as humans do when they are dealing with a large list or set of discrete elements?

In order to find out, they played various violations of interactions between families. The researchers found that an apparent rank reversal of families resulted in a much more dramatic response from listeners than a rank reversal between individuals within a family. This implied that baboons not only recognize individual rank and family rank but integrate them into an even higher order of hierarchy.

The baboon findings confirm what Fitch and other researchers speak about when they refer to the big gap that exists for all animals between what is comprehended and what is produced. Despite the baboons' limited vocal set, they have what Seyfarth calls an almost open-ended ability to learn the sound-meaning pairs of their own species and of other species. When they hear a vocalization, he says, they form a mental representation of each call's meaning. This response seems to be instantaneous and sub-conscious, and it also appears to be an innate property of the baboon mind. Seyfarth suggests that if you are looking for a cognitive foundation that may serve as a precursor to syntax, it's much more likely to be found on the interpretation side than on the production side of animal communication.

It may be that before our ancestors became adept at understanding and producing the computations of modern grammar, they learned to compute social relationships just like the baboon understanding of social rank, which is based on discrete values—individual rank and family rank—and their combination. Seyfarth stresses that this "language of thought" is not the same as human language, but he adds that it is adaptive in its own right and is a possible foundation for something that might turn into language.

Even if animals can understand structural rules where words or cries are joined one after the other, as with the Diana monkeys, human language uses a variety of syntactic mechanisms to build meaning. Thus, while some research has turned up evidence of rudimentary structural abilities in other animals, evidence has also been gathering regarding grammatical rules they are unable to use. Tecumseh Fitch and Marc Hauser tested the ability of tamarins, with whom we last had a common ancestor forty-five million years ago, to understand two different types of grammar.

Fitch and Hauser played recordings of different sequences of sounds to the monkeys. The sequences generated by the first type of grammar could be described by the grammatical rule: $(AB)^n$, where a syllable (A) was always followed by another syllable (B) for a number of (n) repetitions. The sequences generated by the second type could be described by: $A^n B^n$, where the same number of A syllables was always followed by the same number of B syllables. Understanding how the sounds were arranged, according to Fitch and Hauser, required the ability to process two different kinds of grammar, a finite state grammar and a phrase structure grammar. The latter has more expressive power than the former, and it's thought that you can't generate all the structures in human language without at least a phrase structure grammar.

The researchers found that after the tamarins were played the recordings of the first rule, they would react if then played recordings that violated the same syntactic rule—suggesting that they had an expectation about how the sounds would be arranged. However, when the animals were played the sound sequences generated by the second rule, they didn't show any sign that they could distinguish examples of correct syntax from sequences that violated the structural rule—it was all the same to them. Human subjects, in contrast, noticed the violations of both the finite state grammar as well as the phrase structure grammar.

In a interesting demonstration of the tangles created by homology and analogy, Timothy Gentner, an assistant professor of psychology at the University of California, San Diego, and colleagues demonstrated in 2006 that starlings can actually distinguish correct instances of the grammar based on Fitch and Hauser's example, $A^n B^n$. The researchers used natural starling sounds to test the birds, exposing their subjects to many more examples than Fitch and Hauser exposed the monkeys. Gentner and colleagues suggest that these results show the comparative syntactic abilities in monkeys, humans, and birds may differ more in quantity than in quality. So rather than a singular syntactic ability that is a key foundation for human language, there may be a fundamental set of structural mechanisms that we use—some of which other animals also possess.

The Gentner paper received a lot of public attention. Many researchers were surprised by the results, and some welcomed the findings as proof that the syntax underlying human language is not a monolithic ability that only we possess. But the experiment was not universally acclaimed; in a *New York Times* interview Chomsky said that what the starlings did has nothing to do with language at all.

Certainly, the Fitch and Hauser and the Gentner experiments raised many interesting issues about methodology, as well as the capacity for understanding different kinds of grammar. Ray Jackendoff and colleagues published a letter noting that what the starlings are habituating to depends on how they encode the signal. They also questioned whether the starlings were really doing syntax as opposed to basically counting the strings of A's and B's (echoing Chomsky's comment). Recall that many animals have some number ability. Indeed, it's possible that the humans in the original experiment may have been counting the experimental stimuli rather than processing them as samples of a phrase structure grammar. Jackendoff explained: "If I imagine the stimuli and how I would think of them or remember them, it would be by counting or some equivalent rhythmic procedure: matching the first A with the first B, the second A with the second B, and so on. It would not be by relating the first A to the last B, the second A to the next to last B, which is what the syntactic hypothesis entails."

Despite the complications, these experiments inaugurate a potentially rewarding endeavor that seeks to map which syntactic strategies are available to some species and not others.

It's hard to overestimate the intricacy and power of each language's syntax, let alone all of the syntactic strategies that human languages deploy. The complexities of linguistic structure that, so far, do not seem to have an analog in any kind of animal communication are myriad, including many different mechanisms for combining words and parts of words into a larger phrase, and larger phrases into even larger ones. For instance, a phrase of any length may be created by conjoining elements (He knows and she knows) or by arranging them recursively (He knows that she knows).

Parts of a phrase—for instance, the subject and the verb—may agree in number and gender. A verb may be intransitive, taking a subject but not a direct object (She slept) or transitive, taking a subject and direct object (She kicked it). A language may be ergative, marking the object of a transitive verb in the same way that it marks the subject of an intransitive verb, while the subject of a transitive verb is marked differently. Or a language may be nominative-accusative, marking the subject of a transitive and intransitive verb in the same way, distinct from the object of the transitive verb. Different languages mark these relationships in very different ways, using strategies like word order or lexical marking. And often the way a particular language deploys one kind of syntactic rule affects how it fulfills another. For instance, languages with free word order have many syntactically meaningful affixes.

Human syntax is also characterized by countless idioms and phrases, some of which are complete sentences (The jig is up), while others allow single words to be inserted into "slots" to create different meanings, such as "Take X for granted." In yet another type of English idiom (also called a "syntactic nut"),[4] the phrase that complements the verb isn't actually determined by the object of the verb; for example, "He sang/drank/slept/ laughed his head off" or "Sarah slept/drank/sang/laughed the whole afternoon away."[5]

In contemporary syntax there are two main approaches to accounting for all the structural rules that human languages use to build meaning: the Chomskyan approach and the "parallel architecture" approach. In the Chomskyan approach, the list of words in a language—its lexicon—and the syntactic rules that arrange those words are treated as separate entities. Syntactic rules are considered the core computational device with which most of language is generated. Accordingly, people still talk about universal grammar, or UG, as a language-specific set of rules or parameters belonging to a language-specific mental mechanism.

Mark Baker's book *The Atoms of Language* (2001) is a good example of the mainstream approach to syntax. Baker's goal was to show that apparently very different languages, like English and Mohawk, are different only in the way that a finite set of universal rules is applied to create them. Baker deduced a hierarchical list of fourteen parameters that he believes reflect rules that are hardwired into the human brain. He thinks there may be about thirty rules overall. English and Mohawk differ only, he says, in the way one single rule is applied at the top of the hierarchy.

Jackendoff calls this kind of approach "syntactocentrism," meaning that syntax is regarded as the fundamental element of language. In contrast, he says, "in a number of different quarters, another approach has been emerging in distinction to Chomsky's." In this new way of accounting for structure in language, words and phrases are as important as the rules that combine them, and the idea of pure syntax is downplayed.

Instead of being objects, words are best thought of as interfaces. A word lies at the intersection of a number of systems—the sound of the word (phonology), syntactic structure (the structures that the word can license or appear in), and meaning (some of which may be specific to language, and some of which may be a more general kind of meaning).[6] The more general component of a word's meaning may have some equivalence to the common cognitive platform that humans share with other species.

Jackendoff may be the only longtime generative linguist who willingly concedes that we may share this component of a word with a number of other species. As he explains: "An account of the extraordinarily complex behavior of primates, especially apes and especially in the social domain, leads inexorably to the conclusion that they are genuinely thinking thoughts of rich combinatorial structure, not as rich as human thought to be sure, but still combinatorial."[7]

It's significant that Jackendoff now proposes that it's time to move away from the pure focus on syntactic structure and the idea of a syntactic core to language. While he believes that language is as complicated and ramified as Chomsky does, he is now convinced there is a different way to account for that richness.

Rather than think of syntax as a set of computational algorithms, Jackendoff and Pinker call it a "sophisticated accounting system" for tracking the various layers of relationship and meaning that can be encoded into speech and then decoded by a listener. To their mind syntax is "a solution to the basic problem of language," which is that meaning is multidimensional but can be expressed only in a linear fashion, because speech unfolds sequentially, as does writing. This way of looking at language and syntax is more consistent with the idea of language evolution and the view of evolution as a "tinkerer."

Coming from a slightly different viewpoint, John McWhorter, a former professor of linguistics at the University of California, Berkeley, and senior fellow at the Manhattan Institute, emphasizes the way that, like biological evolution, language change results from in accretions or accumulations of structure. In this sense language is an artifact of the collective mind of history. It has imperfections and odd quirks, and makes peculiar demands of its speakers. Its textures and patterns have been created over a long period of time as it has been dragged through millions of mouths, expressing their individual agendas.

McWhorter argues that a lot of syntactic structure is sludge and is not shaped by logical necessity or innate mental rules. He talks about the "benign overgrowth" of language as a corrective to the idea that languages are a lens onto the human mind. He wrote: "There are few better examples than Fula of West Africa of how astoundingly baroque, arbitrary and utterly useless to communication a language's grammar can become over the millennia and yet still be passed on intact to innocent children." Fula, McWhorter points out, has sixteen genders, and each noun gender marker varies according to the noun. Moreover, any adjectives modifying a noun also must carry a different gender marker in order to agree with the noun.[8]

In *Simpler Syntax,* a book coauthored with Peter Culicover, Jackendoff writes that while it is important to ask how optimal or perfect a language is, it is also necessary to recognize that language doesn't operate like a physical system, say, "a galaxy or a molecule . . . It has to be carried in the brain, and its structure has to come from somewhere."

Jackendoff and Culicover conclude by noting that they have heard it said in certain circles that if their ideas about language are true, then it means language is "not interesting." But interestingness, they reply, is in the eyes of the observer. "Baseball cards and poodles interest some people and not others," they write, "and the same is true of simpler syntax."[9]

What can the structure of language itself tell us about the way language changes over time? Linguists have developed a number of ways of investigating this topic.

For many decades linguists have been rebuilding dead languages and tracing relationships between modern languages by using what is called the comparative method. This approach works by matching words with the same (or similar) meaning from different languages. Such words are called cognates; for example, "cow," "swine," and "sheep" in English are cognate with *Kuh, Schwein,* and *Schaf* in German. Linguists have charted details of the relationships between the 150 languages in the Indo-European family.[10] By tracking the way the sounds of words change over time, one can determine not just relatedness between languages but what the original grand language, from which all the modern versions descended, sounded like. In this way, they can plot the successive changes that transformed a language like Old High German into a group of languages like English, Dutch, and Modern German. The comparative method can also be used to demonstrate how the appearance of a relationship can be illusory. For example, one of the Pama-Nyungan languages of aboriginal Australia and Modern English both use the word "dog" to designate the friendly quadruped. Historical linguistic research can establish that the modern-day languages Erzya and Moksha descend from proto-Mordvin, and that proto-Mordvin and proto-Samoyed are variants in time of an even older eastern European language. Linguists believe they can reconstruct around five thousand years of language using this approach. (Some claim they can go as far back as eight to ten thousand years.)

Very recently, another method has been proposed that promises to push our knowledge a little further back in time. Instead of comparing words, this technique uses computer-based strategies from evolutionary biology and compares syntactic structures, such as the ordering of verbs and subjects in a sentence. Researchers at the Max Planck Institute for Psycholinguistics in the Netherlands first tested it against a family of Austronesian languages whose relationships had already been uncovered by the comparative method. They came up with the same results. They then looked at fifteen Papuan languages that were not known to be related. Most of the words in the Papuan set looked unrelated, so the comparative method was of no help in determining whether any relationships existed. Using the evolutionary biology method, the researchers were able to reveal relationships between the languages that, crucially, were consistent with their geographical distribution. They suspect that these languages trace back to a common ancestor that was spoken more than ten thousand years ago.

Jackendoff has developed an approach for recovering ancient elements of language that would take us even further back into the past. He believes that language itself carries fossils of earlier forms, allowing us to reverse-engineer it back to an evolutionarily simpler state. Jackendoff was inspired by Derek Bickerton, one of the first linguists to develop the concept that before our current form of language, we must have communicated with a protolanguage, a simpler step on the way to modern words and syntax.

Jackendoff said: "The idea behind it is that there was this stage of 'protolanguage' preceding the stage of modern language. The logic behind figuring out protolanguage is that we can find aspects of modern language that could have served as an effective communication system without the rest of language. What Bickerton's version of protolanguage has that no other animal communication system has is some kind of phonology, so you can build a large vocabulary. In addition it has the symbolic use of words, and it concatenates words to convey meanings that combine the meanings of individual words. What it doesn't have to have is modern syntax."

Even to achieve this level of protolanguage, you must have two or three very important innovations in place. "The construction-based view of language," Jackendoff explained, "makes it natural to conceive of syntax as having evolved subsequent to two other important aspects of language: the symbolic use of utterances and the evolution of phonological structure as a way of digitizing words for reliability and massive expansion of vocabulary." Once you have that, the rest can follow.

In his book *The Symbolic Species,* Terrence Deacon proposes that various platforms of understanding are necessary for an animal to use utterances symbolically. He invokes three types of reference described by Charles Sanders Peirce—iconic reference, indexical reference, and symbolic reference. Crucially, these distinctions are not inherent to any object or event in the world, but rather are descriptions of the kinds of interpretations that can be made about objects or events. Icons (or making an iconic interpretation) are the simplest type of reference. If an object is iconic, there is a similarity between it and something else. Landscape paintings, Deacon points out, are iconic of the landscape they depict. Indexes are a step more complicated than icons because they are built from iconic relationships. With an indexical interpretation, there is some kind of correlation, often causal, between an object or an event and something else. A skunk smell may indicate that a skunk is nearby. "Most forms of animal communication," writes Deacon, "have this quality, from pheromonal odors (that indicate an animal's physiological state or proximity) to alarm calls (that indicate the presence of a dangerous predator)."[11] A symbol, in turn, is more complicated than an index, because it involves some kind of convention or system that guides the way we link one thing to another. A wedding ring is a symbol of marriage, writes Deacon, just as *e* is a symbol of a sound that we use in speech.

Complicated reference is thus created by layering simpler forms of reference together. Much animal communication makes extensive use of iconic and indexical reference, but only human language is rooted in the unusual and complicated relationships that exist with symbolic reference. The jump to symbolic reference from indexical reference is not straight-forward, argues Deacon. Symbols do not exist by themselves; they exist only in the context of other symbols and, crucially, in the relationships between them. Making a symbolic interpretation involves simultaneously understanding where a symbol, be it a wedding ring or a word, exists with respect to other symbols in its set (in the case of a word, knowing which other words it can be combined with and which it can't) and understanding the way it refers to objects or meaning in the world. Only because we understand symbolic reference and the ways that words must be combined with other words (which is to say that words are by their very nature syntactic) can we create modern human language with all its various structural possibilities. Even though symbolic reference is highly unusual in the nonhuman world, it's not impossible for some animals to comprehend it. Kanzi is a good example of an animal that has been bootstrapped into this sophisticated form of understanding.

In the same way that Deacon lays out the progression of meaning as a logically layered hierarchy, Jackendoff proposes other elements that must have necessarily followed one another in the increasing elaboration of human language over time. He suggests that the next stage after the ability to use symbols was reached must have been the employment of symbols in nonspecific situations. Words like "damn it," "ouch," and "wow" have no syntax at all and are, he says, fossils of this stage. Remnants of the next stage—where symbols are a little more bound by syntax and a little more tied to context—are "ssh," "psst," "hello," "goodbye," "yes," and "no." (No animal communication system has an equivalent for "no," argues Jackendoff, but it tends to be one of the first words in any child's repertoire.)

After these early stages comes the development of a large set of symbols that is, in principle, unlimited. Next is the ability to place these items together in meaningful ways. Syllables and phonemes must have come next, preparing the way for the development of a simple protolanguage. Then, along with the appearance of grammatical categories and inflections, there must have been a proliferation of symbols for encoding abstract meaning, like words for space relationships ("up," "to," "behind") and time ("Tuesday," "now," "before").[12] All these layers were required to build modern language, with its complicated syntax and special linguistic meanings, and all must have been in place before the first ancient human language split into the branches that lead to the many modern languages.

"It's totally hypothetical, this reverse engineering," said Jackendoff, "but it's the kind of thing you would investigate across species if you could: You would look for another species that has only a subset of the features, corresponding to an earlier stage. Of course, there aren't any such species for the deeply combinatorial aspects of language. But if we were looking for the evolutionary roots of eyes or toes this is exactly what we would do."

Another domain in which humans use structure with virtuoso abilities is music, which is, like language, one of the species' relatively few universal abilities. Without formal training any individual from any culture has the ability to recognize music and, in some fashion, to make it. Why this should be so is a mystery. After all, music isn't necessary for getting through the day, and if it aids in reproduction, it does so only in highly indirect ways. Scientists have always been intrigued by the connection between music and language. Yet over the years, words and melody have been accorded a vastly different status in the laboratory and the seminar room. While language has long been considered essential to unlocking the mechanisms of human intelligence, music is generally treated as an evolutionary frippery—"auditory cheesecake," as Steven Pinker put it.

But thanks to a decade-long wave of neuroscientific research, attitudes have been changing. A flurry of recent publications suggests that language and music may equally contribute in telling us who we are and where we have come from. It's not surprising to find that many of the researchers engaged in the study of language evolution are also drawn to the evolution of music. Ray Jackendoff (who, in addition to being a linguist, was the principal clarinet of the Civic Symphony Orchestra of Boston for twenty years) and colleague Fred Lerdahl have investigated language and music as cognitive phenomena. Breaking music into its major components (rhythm, the structure of melody and harmony, emotion in music), Jackendoff and Lerdahl sought to identify which elements of music arise from general cognitive processes, which come from processes that are common to music and language, and what, if anything, is peculiar to music.

Their investigation of musical affect is most interesting with regard to these three questions. Clearly, some affect in music derives from a broader set of associations. For example, we are startled by sudden, loud noises, and this applies equally to random noise as to sudden, loud musical outbursts. In addition, some affect appears to draw on a shared understanding of language and music. Jackendoff and Lerdahl point out that large structures in music can be like dramatic arcs in narratives. The slow buildup of tension, a climax, and then denouement can be found in both musical pieces and stories. It may be that both music and language exploit a human predisposition to understand events in terms of tension and resolution. Jackendoff and Lerdahl also suggest that the way people convert music into gesture, whether by dance or in conducting an orchestra, is instinctive and special to music alone. Different kinds of music invoke different kinds of movement; a waltz does not inspire people to march, and vice versa. Even very young children show a sensitivity to this aspect of music when they spontaneously dance. While it's not possible to fully disentangle these aspects of the musical experience from one another, an investigation of the common and unique cognitive bases of music, say the researchers, contributes to its biological profile, which in turn helps track its evolutionary trajectory.

In an article in the *Journal of Neuroscience,* David Schwartz, Catherine Howe, and Dale Purves of Duke University investigated the question of the language and music relationship in a very different way, concluding that the sounds of music and the sounds of language are intricately connected.[13]

To grasp the originality of their idea, two things about how music has traditionally been interpreted must be understood. First, musicologists have long emphasized that while each culture stamps a special identity onto its music, music itself has some universal qualities. For example, in virtually all cultures sound is divided into some or all of the twelve intervals that make up the chromatic scale—that is, the scale represented by the keys on a piano. For centuries, observers have attributed this preference for certain combinations of tones to the mathematical properties of sound itself.

Some twenty-five hundred years ago Pythagoras was the first to note a direct relationship between the harmoniousness of a tone combination and the physical dimensions of the object that produced it. For example, a plucked string will always play an octave lower than a similar string half its size, and a fifth lower than a similar string two-thirds its length. This link between simple ratios and harmony has influenced music theory ever since.

Second, this music-is-math idea is often accompanied by the notion that music, formally speaking at least, exists apart from the world in which it was created. Writing in the *New York Review of Books,* the pianist and critic Charles Rosen discussed the long-standing conviction that while painting and sculpture reproduce at least some aspects of the natural world, and writing describes thoughts and feelings we are all familiar with, music is entirely abstracted from the world in which we live.

Neither idea is correct, according to Schwartz and colleagues. Human musical preferences are fundamentally shaped not by elegant algorithms or ratios but by the messy sounds of real life, and of speech in particular—which in turn is shaped by our evolutionary heritage. Said Schwartz, "The explanation of music, like the explanation of any product of the mind, must be rooted in biology, not in numbers per se."

Schwartz, Howe, and Purves analyzed a vast selection of speech sounds from a variety of languages to determine the underlying patterns common to all utterances. In order to focus only on the raw sound, they discarded all theories about speech and meaning and sliced sentences into random bites. Using a database of over a hundred thousand brief segments of speech, they noted which frequency had the greatest emphasis in each sound. The resulting set of frequencies, they discovered, corresponded closely to the chromatic scale. In short, the building blocks of music are to be found in speech.

"Music, like the visual arts, is rooted in our experience of the natural world," said Schwartz. "It emulates our sound environment in the way that visual arts emulate the visual environment." In music we hear the echo of our basic sound-making instrument—the vocal tract. This explanation for human music is simpler still than Pythagoras's mathematical equations: we like the sounds that are familiar to us—specifically, we like sounds that remind us of us.

This brings up some chicken-or-egg evolutionary questions. It may be that music imitates speech directly, the researchers say, in which case it would seem that language evolved first. It's also conceivable that music came first and language is in effect an imitation of song—that in everyday speech we hit the musical notes we especially like. Alternately, it may be that music imitates the general products of the human sound-making system, which just happen to be mostly speech. "We can't know this," says Schwartz. "What we do know is that they both come from the same system, and it is this that shapes our preferences."

Schwartz's study also casts light on the long-running question of whether animals understand or appreciate music. Despite the apparent abundance of "music" in the natural world—birdsongs, whale songs, wolf howls, synchronized chimpanzee hooting—previous studies have found that many laboratory animals don't show a great affinity for the human variety of music making. Indeed, Marc Hauser and Josh McDermott of Harvard argued in a special music issue of *Nature Neuroscience* that animals don't create or perceive music the way we do.[14] The fact that laboratory animals can show recognition of human tunes is evidence, they say, of shared general features of the auditory system, but not of any specific musical ability.

But what's been played to the animals, Schwartz noted, is human music. If animals have evolved preferences for sound as we have—based on the soundscape in which they live—then their "music" would be fundamentally different from ours. In the same way our scales derive from human utterances, a cat's idea of a good tune would derive from yowls and meows. To demonstrate that animals don't appreciate sounds the way we do, we'd need evidence that they don't respond to "music" constructed from their own sound environment.

Of course, there are many examples of animal music. Fitch (who is also an avid amateur musician, composer, and singer) argues that it is worthwhile to examine these in comparison to human music and language. Fitch examined not just animal song, like birdsong and whale song—which must be learned as we learn to sing and talk—but also examples of animal instrumentation. The best examples of instrument use in nonhuman animals are found in our very close relatives. For dominance displays and in play, chimpanzees drum on trees and other resonant objects, while gorillas drum on their own chests (and occasionally other objects). Sue Savage-Rumbaugh's bonobos have also demonstrated an appreciation of percussion and keyboard playing (recall that they also use keyboardlike machines for linguistic communication). Instrumental music is rare in vertebrates, except for African apes, which includes us, leading Fitch to suggest that the drumming of chimpanzees and gorillas may be evolutionary homologs to human instrumental music.

Fitch has further explored the antecedents of human instrumentation via the divisive issue of Neanderthal flutes. A number of researchers have examined a fossilized cave-bear bone with two holes (and possibly another three damaged holes), attributed to Neanderthals.[15] It has been argued that the object, which is radiocarbon-dated to approximately 43,000 years ago, is a flute. Although the provenance and nature of this bone are still regarded as controversial, Fitch points out that if it was a flute, it dates the origin of human instrumental music to at least the common ancestor of Neanderthals and humans, *Homo heidelbergensis* (see chapter 12), who lived more than 500,000 years ago.

No matter how the connection between language and music is parsed, what is apparent is that our sense of music, even our love for it, is as deeply rooted in our biology as language is. The upshot, said the University of Toronto's Sandra Trehub, who also published a paper in the music issue of *Nature Neuroscience,* is that music may be "more like a necessity than the pleasure cocktail envisioned by Pinker."

This is most obvious with babies, said Trehub, for whom music and speech are on a continuum. Mothers use musical speech, called motherese, to "regulate infants' emotional states," she explained.[16] Regardless of what language they speak, the voice all mothers use with babies is something between speech and song. This kind of communication "puts the baby in a trance-like state, which may proceed to sleep or extended periods of rapture." This means, explained Trehub, that music may be even more of a necessity than we realize.[17]

10. You have a human brain

On July 28, 2005, Lacy Nissley was scheduled for neurosurgery at Johns Hopkins Hospital in Baltimore. Before she was born, the neurons in Lacy's right hemisphere migrated to the wrong place in her brain. The hemisphere became enlarged and started to cause seizures that were only poorly controlled by medication. As time went on, Lacy's seizures got worse. Nothing could be done to make her right hemisphere work well, and while it was attached to the rest of her brain, it corrupted the way the left hemisphere worked. The only chance Lacy had to live a normal life was for her to undergo a hemispherectomy. In this radical operation, Lacy's neurosurgeon would remove her right hemisphere, essentially taking out half of her brain.

Four hours into the operation, Lacy's neurosurgeon, Dr. George Jallo, his resident Dr. Violette Renard, and the OR nurse Sean Stelfox stood in a small, still crescent around Lacy's head. Earlier, Jallo had removed the frontal lobe. He then used micro-scissors to cut around the parietal lobe, and now he and Renard were slowly working their way around each side, making tiny little pinches into the cut with electric cauterizing forceps. Occasionally, Jallo used a flat metal spatula to lift the lobe up and back so he could push the bipolar forceps farther in. As the cut became deeper and wider, the tissue on either side browned and blackened, and the lobe, which was initially stationary, started to move back and forth as more of it was detached from the rest of the brain.

Deep at the bottom of the parietal wedge lay the white matter of Lacy's brain. Everything else was colored or discolored, but the long cables that connect neurons to one another gleamed toothpaste white. They came apart like string cheese. Stelfox bent toward Jallo clutching a small plastic

bowl with both hands. Using normal forceps, Jallo picked out the lobe—
it was the size of an infant's fist—and dropped it into the container. Stelfox
held it aloft. "The parietal lobe."

Four hours later, the right hemisphere was gone.[1] From the top of
Lacy's head, her cranium looked like a wide, uneven bowl, revealing the
white-pink base of the skull from the inside and the larger, deeper cavity
that had held the frontal and parietal lobes. In the middle was a shallow
mound where Jallo left a layer of axons to protect the ventricle. The white
matter was now gray-black. Jallo and Renard lightly touched their forceps
to it, and the cauterizers fizzed, and occasionally popped and spluttered,
sealing the brain against micro-hemorrhages. Just below the mound were
the basal ganglia, small dark squiggles in the emptiness. Over and over
Stelfox poured in saline, and Jallo and Renard drew it out again.

Jallo filled the right side of Lacy's head with saline, and over the next
few days it would be replaced by the brain's constant drip of cerebral spinal
fluid. He then reattached her skull using four tiny dissolvable plates made
of sugar. Overall, the hemispherectomy took nine hours, and at the very
end Renard bandaged Lacy's head and gently turned her onto her right
side, sticking on tape that said "This side up."

Lacy was released from the hospital a week later.[2] Around one hundred
children have undergone a similar procedure at Johns Hopkins, and with
extensive therapy to help them relearn how to walk, talk, and think, the
overwhelming majority of them have flourished.

Hemispherectomies are a drastic but necessary operation for a small
group of people, most of them children. Faraneh Vargha-Khadem, a professor of Developmental Cognitive Neuroscience at the University College London Institute of Child Health, has followed up on a large number
of children who have undergone hemispherectomy. Her best-known case
was Alex, a young boy whose left hemisphere was removed when he was
eight and a half years old. Alex was virtually mute before the surgery, and
his comprehension of words had developed only to the level of a four-year-
old. But around ten months after his left hemisphere was taken out and his
antiseizure medication was withdrawn, he began to speak first in single
words and later in phrases and then in sentences. Even in the normally dry

tones of science journals, you can perceive the researchers' surprise. "To our knowledge," they wrote, "no previously reported child has acquired a first spoken language that is clearly articulated, well-structured and appropriate after the age of six years."

How can a brain do such a thing? At this point in human evolution, there are so many neurons in our brains that the potential number of connections between them is thought to be around 500 trillion. We've had these enormous brains for about 200,000 years, and it took us almost all this time (190,000 years) to start opening our skulls and interfering with them. It took another 9,900 years to really start working out how the brain functions. Since 1990 the neuroscience of language has run a course similar to that of animal cognition and language evolution in that it has undergone revolutionary changes. Our picture of language in the brain since then has been transformed almost beyond recognition.

Nothing in the traditional view of how the brain and language function could account for Lacy and Alex. A skeptic might argue that Lacy can talk because her right hemisphere was removed—scientists used to believe that language was located almost entirely on the left side of the brain. But if that were the case, Alex would be forever mute. Indeed, in the last few decades, a number of children have demonstrated that they are able to talk after removal of the left hemisphere. Most of them suffer some kind of deficit, but their language is more than good enough to enable them to get by in the world.

In the past the only way to deduce the workings of the brain was through the successes and mistakes of primitive neurosurgery and "experiments of nature," cases where unfortunate individuals suffered brain damage from some kind of accident. Observers were able to determine the damage postmortem and then plot in a crude way how it had affected behavior and thinking while the victim was alive.

Phineas Gage is the best-known case study in accidental neuroscience. Gage was a railroad laborer, and in 1848 the inadvertent sparking of some gunpowder sent a bolt of iron shooting through his brain. He survived, but his personality changed completely. He became surly and difficult and struggled with decision making and planning. Gage's state before and after his injury revealed a great deal about the role of the frontal lobes in the workings of the brain.

Today magnetic resonance imaging and positron-emission tomography allow scientists to peer inside a normal living brain and see how it works in real time. Electroencephalograms, another useful technology, measure the electrical waves that are naturally emitted by the brain. These brain waves change in response to different input, which in a language experiment might include normal and ungrammatical sentences. More recently, neuroscientists have developed a way to keep neurons alive for days at a time in petri dishes. The researchers stimulate the neurons in different ways and watch how they respond.

In the traditional phrenological model, different talents and tendencies existed within separate compartments of the brain, and for a long time people assumed that much of the evidence from brain damage suggested that language existed within specific spaces. But as knowledge about the workings of the brain accumulated, the idea that only one particular part of it was devoted to language progressively weakened and finally was rejected. No neuroscientist has found any specific area or tissue that controls language and language only. There are no obvious neural add-ons in the human brain, and of all its cell types there isn't one that only humans have.

As recently as twenty years ago it was taught that language specifically resided in Broca's and Wernicke's areas on the left side of the brain. It's hard to even imagine now how confidently that belief was held, because as we know today, language function is spread throughout the brain. According to Fred Dick, a senior lecturer in psychology at Birkbeck, University of London, all the laboratories that have tried to find a language area have been successful in that they have indeed found dozens, even hundreds, of them.[3]

If you look for activation in any cortex, when language is spoken or comprehended, you will find it. Lieberman's studies of Parkinson's patients and Everest climbers, as well as Pinker's work on the past tense in English, show that there is an overlap between the parts of the brain that are used for speech and the parts that are used for syntax. In addition, the brain areas that are active when learning language are different from the ones that are active when using language once it has been learned. More-over, different areas are activated depending on the specific language ac-tivity, like the comprehension of words, categorizing a word (in a new task versus a learned task), translating between languages, or making decisions about grammar.[4] Modern brain imaging has also revealed that the spread of language activation across the two hemispheres of the brain can differ substantially for each individual.[5]

Clearly, there is no one-to-one correspondence between an area in the brain and all language ability. Although the brain does contain identifiable areas, complicated behaviors are underwritten by many different groups of neurons, and these are linked together to form circuits.[6] The activation that takes place within a small, identifiable part of the brain is often a part of a much larger circuit of activation that is distributed throughout the brain. Walking, striking a piano key, speaking, and listening to speech arise from these large neural circuits.[7]

Summing up our understanding in 2002, Lieberman wrote: "Although our knowledge is at best incomplete, it is clear that many other cortical ar-eas [other than Broca's and Wernicke's] and subcortical structures form part of the neural circuits implicated in the lexicon, speech production and perception and syntax, and the acquisition of the motor and cognitive pat-tern generators that underlie speech production and syntax." He lists the cerebellum, the prefrontal cortex, frontal regions of the cortex, posterior cortical regions, the anterior cingulate cortex, and regions of the brain traditionally associated with visual perception and motor control."[8]

The belief that language was located in the left hemisphere was based primarily on the fact that when people suffered damage to Broca's area, the aphasia they experienced appeared to destroy a lot of grammatical knowl-edge. But the data are inconclusive, and as Elizabeth Bates[9] and Fred Dick have pointed out, people with Broca's aphasia are still able to make certain types of grammatical judgments.[10] In fact, it seems they retain a great deal of knowledge of their language's grammar, but have trouble accessing it. Moreover, the symptoms of Broca's aphasia have also been reported in other groups who do not have damage in that part of the brain. Dick adds that the problems that Broca's patients have can be language-specific (though much of the original testing for Broca's was done in English, the

findings were thought to be true regardless of which language and syntactic system the subject used). While this doesn't mean that Broca's area
isn't important for language, it does show that it isn't the only language-
involved area of the brain.

Not only are language and other higher mental abilities distributed
throughout the brain, but Broca's area has been shown to serve other functions as well.[11] As Bates and Dick note, "Activation in Broca's area is observed when subjects plan covert nonspeech mouth movements, make
rhythmic judgments, or perform complex sequences with the hands and
fingers. In fact Broca's area is active when the subject merely observes such
movements by another human being or reacts to static objects (tools) that
are associated with such movements."[12]

None of this evidence against the language-is-a-box-in-the-brain model
means that language is just a function of a homogeneous general intelligence. Bates explained: "There is no such thing as vanilla cognition . . .
There are variations in computational style and computational power from
one region to another, from one layer to another within a single region,
and from cell to cell."[13]

Also, once a human brain has matured, the distribution of language
functions across that brain is not random. Particular areas take on important parts of the overall task of perceiving and understanding language. It
is widely accepted that different sides of the brain dominate in the processing of prosody (right hemisphere) versus syntax (left hemisphere). In 2005
Lorraine Tyler and colleagues published an experiment that compared the
perception of verbs that were regular ("jump," "jumped") and irregular
("think," "thought") in their past-tense form. They demonstrated how the
sound, meaning, and structure of a word all appear to be processed in different areas of the brain.

Brain imaging showed that in the experimental subjects regular past-
tense forms are processed by a neural circuit that includes the left superior
temporal gyrus, Wernicke's area, and connections to the left inferior
frontal cortex.[14] Irregular verbs, however, take a different path through
the brain. It appears as if the stem and affix of the regular past-tense verbs
are computed as the words are heard, but the irregulars, which have no
special syntactic marking, are treated simply as whole words, like nouns or

uninflected verbs. Accordingly, people who suffer brain damage have been shown to have trouble with one type of past-tense verb or the other—but not necessarily both.[15] Fine-grained brain imaging reveals that even if parts of the brain, like Broca's area, perform many nonlanguage functions, they may still be very important for specifically linguistic processing.[16] Such findings underline yet again the way that what we experience as a single thing—language, words, tense—arises from an amalgam of more and less general strategies.[17]

Dismissing the principles of phrenology doesn't rule out the possibility that human children are born with some specialization for language. Those with particular types of brain damage do experience delays in acquiring language. The fact that these children are slowed down suggests that the damaged areas may have been particularly fertile ground for language acquisition before the damage. However, the same children often naturally catch up to a normal level of language use, also suggesting that there are mechanisms that help the brain to recover, to reorganize on the fly. So even if there are parts of the brain that are best suited for language acquisition from birth, other areas can sometimes step in if they fail. The way that a brain can take different routes to the same basic behavior—in this instance, turning language loss into language gain—is called plasticity.

Brad Schlaggar, a pediatric neurologist and a professor at Washington University in St. Louis, says that the best way to think of plasticity is as a support structure. When he gives a talk about plasticity, he always shows students slides of the St. Louis Arch. "As the structure goes up," he explains, "the relationship between the scaffolding and the leading edge of the two sides of the arch changes as they rise up to meet in the middle. The relationship between the scaffold and the emerging mature structure is dynamic, as opposed to a scaffold that surrounds a building and then comes down again." So if damage occurs to the brain of a seven-year-old child, it occurs in a completely different context than if the child were much older or younger. "The scaffolding idea means that even in adults, the organization of the brain for learning a novel task or a challenging task is different from the organization of implementing that task once you have acquired the skill." The scaffolding for language seems to be particularly flexible.

Fred Dick describes the development of language as a moving target. If damage is sustained in one area, language may move, morph, and settle into another.

In his doctoral work Schlaggar transplanted the visual cortex of one fetal rat brain into another, placing it in the spot where the somatosensory cortex, which normally controls the body as it moves through space, typically develops. Schlaggar found that the transplanted visual cortex grew into a fully functioning somatosensory cortex. The inputs into the new region came from the body as it moved in space, and as a result that neural tissue became wired to process that kind of information.

We tend to think of the brain as developing on a completely separate trajectory from that of the body. Traditionally researchers imagined that the brain had some kind of central developmental controller instructing different parts to assume responsibility for different abilities (the visual cortex develops particular types of neurons, while the auditory cortex develops differently specialized neurons, and so on). But recent research has cast grave doubts on the existence of any kind of central controller. It looks as if the brain tissue that ends up becoming part of different specialized regions is not necessarily fated to end up that way, and that input to the brain coming through the filter of the body contributes to its architecture.

Leah Krubitzer, a professor of psychology at the University of California, Davis, also demonstrated how the immature brain isn't fated to be mapped into the specific regions that are typical of the adult brain. She removed a big chunk of the brain of newborn marsupials, and then let them grow up and develop normally. After they reached adulthood, she took another look at their brains. The cortices had organized themselves into exactly the same areas as a normal brain would, all in the same spots relative to each other, but they were all slightly smaller, so as to fit within the smaller brain. While there is a default optimum map, it appears that the map can be drawn over different kinds of neural terrain.

For all the apparent complexity of the human language-brain relationship, it's important not to lose sight of the fact that some hard-to-pin-down behaviors and preferences appear to be completely controlled by the way genes have built the brain. In 2001, in a strange complement to the experiment in which chickens with transplanted bits of quail brain ended up producing some species-specific quail calls, Evan Balaban and colleagues at the Neurosciences Institute in San Diego transplanted a piece of brain from a Japanese quail into the brain of a domestic chicken, and likewise placed a piece of chicken brain into the head of a Japanese quail. With their new chimeric brains, the birds continued to produce the calls of their own species, but instead of responding to the maternal calls of their own species, they showed interest in the calls of the other.[18] There's no reason to believe that processes like these aren't also relevant to the human experience, even if they can't fully explain the complexity of language.

Plasticity means that the early specialization of human brain tissue does not have to be its ultimate destiny. It's more like a career path, with the potential for a future change of jobs. This flexibility applies not just in what the brain can do but in how it is organized. Indeed, plasticity is the way our brain responds to all learning and experience during every minute of every day, regardless of whether that experience is an Italian class or brain surgery. There is no little field linguist inside our heads dividing language up the way we do it consciously; rather, we are plastic, and with plasticity, the hardware is the software.

Plasticity is not just a human trait. In pioneering work, William Greenough at the University of California, Los Angeles, showed that the dendrites and synapses of rats and hamsters change when the creatures are placed in a stimulating environment, and in 2005 a group of Princeton researchers demonstrated that when marmoset monkeys were moved from a standard laboratory setting to a more complex, enriched environment, their dendrites and synapses likewise underwent a dramatic change. The researchers concluded that the primate brain is extremely sensitive to even small increases in environmental complexity.[19] Sue Savage-Rumbaugh invokes plasticity to explain Kanzi's and Panbanisha's extraordinary abilities, especially in comparison to Tamuli, who was exposed to language much later in life and never really acquired it. "By being immersed in a symbol-using environment during the period of greatest brain plasticity, all the components necessary for language comprehension (and production) were put into place for Kanzi and Panbanisha," Savage-Rumbaugh wrote. If the bonobos are exposed to linguistic information at this crucial stage, it appears that their brains can adapt and organize in such a way that they can participate in human culture, even if it's only at the level of a child.[20]

In a 1991 article about Kanzi, Chomsky was quoted as saying, "If an animal had a capacity as biologically sophisticated as language but somehow hadn't used it until now, it would be an evolutionary miracle."[21] Yet it's clear by now that many surprising and sophisticated capacities can be acquired by individual animals that they do not necessarily use in the wild. Plasticity suggests that mental variety is a fundamental characteristic of animal life, and that different environments can elicit different brains and mental skills from the creatures within a single species. A pathologist's examination of brains from language-trained apes may help illuminate the specific changes that language seems to induce in the plastic brain. So far only one such organ has been examined. It weighed 528 grams, much more than that of the typical chimpanzee brain.[22]

The ideas of Schlaggar, Dick, Krubitzer, and other researchers are generations away from the search for the one or two nuggets of difference between speaking humans and nonspeaking animals. Carving up the world into qualitative differences may make sense to us psychologically, but it is not supported by biological research. Language as a whole is a phenomenal mental and social skill, but the enormous differences between being able to speak it and not do not correspond to equally large differences in the physiology of the brain.

Lacy's and Alex's recoveries are shocking to us in part because of the deeply held belief that it is the size of our brains that distinguishes our species. For a long time, we have assumed that the sheer bulk of the human brain was what made it such a formidable computing machine. We assumed a simple one-to-one relationship between intelligence and brain size, such that a brain will think more if there is more of it and, accordingly, it will think less if it is smaller.

But in absolute terms humans don't have the largest brains (whales do). What we have, rather, are the biggest brains with respect to body size of any animal on the planet. The ratio of brain size to body size is called the encephalization quotient, or EQ. This measurement is based on the assumption that you can predict how much brain tissue an animal needs given how large its body is. Any extra tissue over and above that minimum is considered a bonus and a marker of intelligence.

Lori Marino, one of the researchers on the dolphin mirror-image project, investigates the possibility of using EQ as a neutral, objective measure of intelligence across species. She has examined cranial fossils to determine the EQ of dolphins and humans over the course of history. "I'm trying to understand what the big patterns are, and whether those patterns are driven by the same processes in humans and other animals. Fundamentally, all brains operate under the same physical laws. So my view is we should be looking for general principles and then possibly the uniqueness to each group."

Humans currently have the highest EQ of all organisms, about 7. Bottlenose dolphins have a particularly high EQ (4.2), while belugas measure a respectable 2.4. In general, cetaceans—whales, dolphins, porpoises—measure from 1 to 5, chimpanzees measure 2 to 3. New Caledonian crows have a high EQ with respect to other birds. (No one has yet investigated the EQ of insects, or even whether it would be an appropriate measure for this type of organism.)[23]

Encephalization is only half the picture, said Marino. "You can have two brains that are just as big as each other, but organized in different ways, and one can be a much more complex information processor." Comparing EQs over many species is the beginning of a truly non-human-centered approach to measuring brains.

Incidentally, the ranking of animals with the highest EQ has changed a number of times over the last few million years. Mainly, humans have been jockeying for first place with dolphins. Marino points out that a number of dolphin species throughout history had very similar brain-body ratios to our ancestors—*Homo habilis,* around 2 million years ago, and *Homo erectus,* only 1.8–2 million years ago. Rankings have shifted in the blink of an evolutionary eye and perhaps, said Marino, could do so again. "The past couple of million years at most is really the only time in history that humans have been the most encephalized organisms on the planet. It just wasn't so two or three million years ago." Our current standing with the biggest EQ may be secure because our highly developed culture props us in first place.

But then again, our position may not be as strong as we think. On a planet that's been in existence for four billion years, and at a completely arbitrary slice of time, can one species really be certain that things won't change? (Presumably, if they do change, the dolphins will explain to us where we went wrong.)

Terrence Deacon brings together the perspectives of neuroscience, semiotics, and biology in order to examine the ancestral human brain as it enlarged and what effects the changes in brain-body ratio had on our abilities and behavior. He argues that first we need to compare the growth rate of our brains with those of other species. It turns out that human brains are two steps removed from the general growth patterns of all mammals.

First, humans are primates, and at some point in the distant past the primate brain evolved such that it grows a bit differently from all other mammal brains. Indeed, all primates are at least twice as encephalized as other mammals. Generally, we assume that this greater encephalization results from larger primate brains being selected for greater intelligence.

But, Deacon points out, encephalization measures a relationship between brain and body. It's not that the primate brain got bigger, he argues, but that primate bodies started to grow smaller. Deacon compared the body and brain growth rates of primates and other animals. He found that primate brains and other mammal brains grow at the same rate, but that primate bodies grow at a slower rate than other mammal bodies. So while primate brains continue to develop along the same growth trajectory as those of other animals with a similar evolutionary history, primate bodies grow more slowly and therefore, over time, got *relatively* smaller. As primates, our ancestors rode that wave of greater-encephalization-by-smaller-body.

Second, humans changed once again. We are three times as encephalized as other mammals and one and a half times as encephalized as other primates. This is due to the fact that our brains not only grow at the typical primate rate but grow for longer. At the point that other primate brains stop developing, human brains continue to do so, and for a significantly longer period of time.

The mismatch between the growth rate of body and brain in humans as compared with the mammalian average can best be understood by imagining what we'd look like if our bodies grew at the same rate as our brains, says Deacon. If our body and brain growth rates matched, humans would look more like *Gigantopithecus,* a half-ton Asian ape that became extinct in the last few hundred thousand years.[24]

The work of Marino and Deacon emphasizes how important it is to take subtle and complicated relationships into account when we make comparisons across species. Simply taking the brain of one species and comparing its gross size with another's, is, in the end, not going to answer many questions about why one brain can support a vocabulary of sixty thousand words and complicated syntax, while the other cannot. Other researchers in recent years have uncovered important commonalities in animal brain anatomy and in the function of various types of neurons.

Evidence of the ancient neurological connections between language and gesture were announced in *Nature* in 2001, when Claudio Cantalupo and William D. Hopkins found that a crucial part of the brain that has been linked with language in humans, Brodmann's area 44, which is part of Broca's area, exists in chimpanzees and gorillas as well. What was striking about this discovery was not merely the existence of the area in other primates but the similarity of its structure to that of humans.[25]

It's common knowledge that the brain is divided into two hemispheres, each of which normally controls the opposite side of the body. Roughly speaking, the right side of the brain controls the left hand and leg, and vice versa. It's also the case that particular functions and behaviors can be dominant in one hemisphere, and a significant amount of language function seems to be represented on the left side. Brodmann's area 44 is larger on the left side than on the right in humans. So far so good—we've known this for a long time. But Cantalupo and Hopkins showed that the area corresponding to Brodmann's area in ape brains is much larger on the left side as well.

Why would this be the case? Apes don't speak. And if spoken language is a purely human phenomenon, this finding makes no sense. It does make sense, however, if we think of linguistic ability as having a heterogeneous structure. If this ability has developed piecemeal over time, then ape brains should share some of the same structures we use for language. The ape asymmetry also means, wrote Cantalupo and Hopkins, "that the neuro-anatomical substrates for left-hemisphere dominance for language were evident at least five million years ago and are not unique to human evolution."

But still, apes don't speak. What purpose does a larger left area 44 serve for them? Cantalupo and Hopkins suggest that apes are controlling gestures with this part of the brain in a languagelike way. Humans evolved the ability to point intentionally with their body parts and then with words. Captive apes are known to point at objects with intention, and in the apes observed by Cantalupo and Hopkins, a preference was exhibited for doing so with the right hand. Since the right hand is controlled by the left hemisphere, Brodmann's area 44 may be controlling the ability to flexibly refer to objects in the world, an ability that underpins verbal and gestural communication.

It is also the case that the apes' bias for using the right hand was consistently greater when they were vocalizing at the same time as they were pointing. In evolutionary terms, say the researchers, this means the "brain area may be associated with the production of gestures accompanied by vocalizations." So what started out as a meaningful gesture plus screech in apes, according to Hopkins and Cantalupo, likely became selected over time for speech and modern language in the human species. In 2002 Elizabeth Bates and Fred Dick reviewed the work done on gesture and language and found that as a child grows, these components develop at the same time in the same places in the brain.[26]

Another extremely striking finding about these shared brain bases came from Michael Arbib and Giacomo Rizzolatti, who discussed mirror neurons as the first real evidence of the neurological underpinnings of imitation in 1997. Mirror neurons are specialized brain cells that fire if you, say, grasp a pen; they also fire when you see someone else grasp a pen. In some sense, the brain interprets these actions as the same thing by mapping them in the same way, meaning that what the monkey can do, the monkey can see. Arbib and Rizzolatti argued that the evolution of minor neurons allowed humans to be skilled imitators: what the human can see, the human can, within reason, do. They help explain why speech is rooted in gestural communication.

Over the course of his career, Arbib's research has involved developing computational models of the brain mechanisms that underlie language and getting them to sync with findings in psychology, philosophy, and linguistics. In the 1980s he began a research program at the University of Southern California for computational modeling of mechanisms in the monkey brain and started collaborating with Giacomo Rizzolatti on how the brain used vision to control hand movements. He was thus on the spot when Rizzolatti's research group in Parma, Italy, discovered mirror neurons. It was this work that led him to mirror neurons in monkeys. Arbib began a collaboration with Scott Grafton, a colleague who was an expert in PET imaging, and together they ran some PET experiments to look for mirror neurons in humans, which they eventually found in many regions of the brain.

At first, mirror neurons were thought to underlie only visual recognition of hand actions. But then Evelyn Kohler and others in Parma began to look at their use in the auditory domain, finding that the monkey mirror system was much more sophisticated than originally thought. Mirror neurons fire when stimulated by distinctive sounds as well. For example, if a monkey sees another cracking a nut, certain neurons will fire. If the monkey only hears the breaking shells, some of the same neurons—the audio-visual mirror neurons—will fire. This is a long way from speech, but it does show that mirror neurons can link to auditory input, so some basic mechanisms for grounding the evolution of speech analysis were, presumably, already in place in the brains of our common ancestor with monkeys, who lived twenty million years ago. One aspect of language for which the mirror system may be responsible is the repetition of pronunciation and words. It may also be a foundation for word acquisition, in which repetition is a relatively stereotyped performance.

In his comparative work on mirror neurons, Arbib said his challenge is to ask, "What is the minimal set of requirements for our brain which would make it possible for us to acquire language?" He uses the slogan "the language-ready brain" to suggest that a brain "might not have language, but might be ready to learn it—just as we have computer-ready brains and today we can use computers." He added, "Nobody would claim that our biology was in any way influenced by the use of computers."

So far most researchers have studied one relatively small local area of the brain. Arbib has examined the interaction of mirror neurons in the neocortex, and he's done a fair bit of work on the basal ganglia, the same area of the brain that fascinates Philip Lieberman. Lieberman argued that the kind of sequencing that the basal ganglia controls is as fundamental to language as it is to dancing. And Arbib is inclined to agree. "The mirror system won't explain all of language," he said. "The next big step is to pull together all these brain areas that are very important for language, and in particular for understanding how language is created and understood on the fly. The brain is a big place."

Currently, Arbib is working on a scene description study. "I'm asking, 'How do you look at a scene, where you do start?' If I give you a video clip, you're going to pay visual attention to it, and you're going to create a visual representation. Then you're reading part of it out as a sentence as you describe it to me." When people do this, there's no sense that they are developing a syntactic structure first and then popping words into it as they describe it, but are literally making it up as they go along. The subjects have a very complex mental picture, and they have to translate from the mental picture to the meaning and then to the words of language.

"It's not just the sequence but the skill," said Arbib. As he reached for a cup of coffee, he said: "Take an example from manual skill. We've actually done models of the cerebellum where we reach for a cup. What you'll see is just one smooth movement where my opposing fingers reach the cup at the same time." But he explained that if the cerebellum was damaged, it would not be so easy:

> You'd have to decompose it in two movements, because you can't coordinate the timing. So if you tried to do it, you might end up having the fingers too close when you hit the cup, or too far apart when you reach the cup. In other words, you run the risk of knocking the cup over. So instead, what do you do? You very quickly compensate for your understanding of your deficit, and you reach out and you get, let's say, thumb contact, and then you will close the hand. In other words, you break the thing down into pieces that you know you can succeed with, and then you resynthesize the sequence that will get you to your

goal. But each gesture is itself less skillful than it would be if you exe-cuted it with [an undamaged] cerebellum.

This implicates yet another part of the brain. "You can get the sequence right without the cerebellum, but if you want a smooth performance, you need a cerebellum. It cues each movement, and it coordinates the movements," he said. "You can't do language without a cerebellum."

11. *Your genes have human mutations*

There is a family in England known in the medical literature as the KE family. Its twenty-nine members are spread over three generations, and fourteen of them have severe difficulties with speech and language, as well as some general cognitive problems that are less severe. Faraneh Vargha-Khadem, the cognitive neuroscientist at the Institute of Child Health in London who has studied the family for over two decades, explains that their disorder causes immobility in the lower portion of the face, including the lips, tongue, and mouth.

As a result their articulation is greatly impaired, but the problem is more than one of just motor control. In simple repetition tests, affected individuals have trouble reproducing sounds and words in the correct sequence, selecting the right sounds for words, and maintaining an appropriate rhythm. Multisyllabic words like "hippopotamus" can be particularly difficult, and in general, the more unfamiliar a word, the more trouble they will have saying it. Their speech is sometimes unintelligible.

As babies, the affected family members behaved somewhat like deaf children—they were quieter than the average infant. Because the lower part of their face was relatively immobile, they had a limited array of facial expressions, which in general were not as spontaneous as those of the unaffected members of the family.

In the affected family members, structural and functional brain scanning shows changes in the speech and language areas. For example, when you'd expect Broca's area to be active, the affected KE members show a scattered pattern of activation in regions of the brain that wouldn't normally be active during language processing.

Vargha-Khadem discovered the family when one of the affected children was seen because of speech and language-related problems. Consequent to this meeting, other members of the family were also assessed, and the profile characteristic of the affected individuals was identified. The disorder, she said, involves a complicated circuit that regulates the movement of the muscles of the lips, tongue, and lower face used in speaking and the hardwiring of the brain structures that are typically used for language. It's unknown whether the problems begin with the physical challenges that the affected family members have in producing, and to a lesser extent comprehending, speech, or whether the fundamental obstacle lies in the creation and understanding of language in the brain.

Vargha-Khadem asked a group of geneticists at the University of Oxford to see if they could identify the defective gene causing the disorder. The team spent several years closing in on the gene responsible when their search was given a boost by a similar speech and language problem in an unrelated child. That child had problems very much like those of the affected KE family members, and between the two different sets of data the geneticists were able to narrow their focus and find the problem gene, dubbed FOXP2. It was the first, and so far only, time that a single gene was linked to an inherited speech disorder.[1]

The FOXP2 gene is located on chromosome 7. Because all normal people have two copies of a chromosome, every individual should have two copies of chromosome 7 and two copies of FOXP2. In the KE family affected members have a mutation that leaves them with only one working copy of FOXP2. The result is what geneticists call a dosage effect: If you have two normally functioning copies of FOXP2, brain and language develop normally, as is the case for the unaffected members of the KE family. If you have only one working copy of FOXP2, you are going to have an array of difficulties with language and speech. No living individual with two malfunctioning copies of FOXP2 has ever been found.

FOXP2 is expressed in several organs of the body, including the brain, where its pattern of expression appears to be specific to regions involved with the development of motor control.

Twin and other developmental studies have demonstrated a strong link between genes and disorders of speech and language, but most of these findings have presented a very complicated picture and it is suspected that language impairment is related to many genes. The KE family is the first clear demonstration of a single gene affecting language ability and speech articulation. It is a landmark case and may yet prove to be the twenty-first century's Phineas Gage in its being a foundational case study for more than a century's worth of neuroscience.

The announcement of the discovery of FOXP2 inspired a debate about what role the gene would normally play in language function—whether its main function is to process and produce the sounds of language or specific parts of language, like grammar. Initially, it was hailed in the popular media as proof of the existence of a language gene, or even a grammar gene.

Even before the discovery of FOXP2 some researchers argued that the KE family proved the existence of a grammar gene. Why is the idea of a specific language or grammar gene appealing? Why would isolating a gene that controls language and that controls only language be such a coup? First, if such a language gene existed, you could track the development of language very finely over time. The beginning of language in the human race could, theoretically, be exactly pinpointed. Second, if a language gene like this was discovered, it would give great weight to the theory that language appeared with a big bang. Finally, a language gene that was possessed by humans and no other animal would provide compelling evidence for the traditional claim that language is a discrete mental trait unique to our species. Indeed, in 1990 the linguist Derek Bickerton proposed that language evolved because of a single genetic mutation.

One criticism of Chomsky's view of evolution was that it was almost creationist and that it required some kind of miraculous genetic big bang. The defense to this criticism had always been that Chomsky's ideas about language didn't implicate evolution one way or the other, and yet Bickerton made explicit what critics said was implicit all along. Bickerton proposed that in a single female who lived approximately 220,000 years ago, a genetic mutation resulted in changes to the vocal tract and skull, as well as a rewiring of the brain for syntax, thus giving rise to language.

Bickerton's proposal was vociferously criticized by evolutionary biologists, and he has since modified his position.[2] FOXP2-based claims for a grammar gene have likewise petered out. Why? They depend on a view of genes only as atomistic building blocks and the genome as a blueprint for the organism, and neither of these ideas has held up. While few researchers would claim that language and genes are not related, there has been little evidence that language is genetically encoded. Certainly, there is no direct relationship between possession of the FOXP2 gene and fully having language.

This thread in the development of the language gene story runs strongly parallel to all the ideas regarding comparative animal work on gesture and cognition that have so far been discussed.[3] The genetic mutation idea echoes all the other suggestions that the extremely complex apparatus that allows you to learn language evolved as a discrete and singular entity—a language organ—that arose without any antecedent.

When Darwin described natural selection a century and a half ago, he was essentially describing a genetic process (the way that genes throw up random mutations and then propagate over time). Today we know that all normal humans have twenty-three pairs of chromosomes, which reside in the nucleus of cells. Chromosomes are made up of DNA, which in turn is made up of four nucleic acid bases, adenine, thymine, guanine, and cytosine. The bases are most commonly designated by their first letters, A, T, G, and C. Stretches of DNA along the chromosome constitute a code for specific proteins, so when molecular machinery reads these segments of DNA, proteins are made in the cell. These segments—these units of code—are called genes. A gene is expressed when the protein it codes for has been produced.

In between the genes, there are stretches of nucleic acid bases that do not code for proteins. These strings of A, T, G, and C, junk DNA, can randomly vary without affecting the organism. The genome of an organism, then, is the entirety of its DNA, junk and genes.

During reproduction, genes are duplicated and carried forward, sometimes having no effect whatsoever. Other times, genes or larger groups of genes get flipped and reinserted in the process of duplication, possibly into the same spot, or they might get moved. This rearrangement occurs at different rates in different species (it is a process we don't fully understand).

In the last few years our ability to describe what the units of evolution look like and do has culminated in the sequencing of the human, mouse, rat, fruit fly, and chimpanzee genome, among others. We have discovered that our genome is not nearly as large as we thought, and once we got over the shock of this, our understanding of how genes actually work has grown immeasurably more sophisticated. The sense that a huge gap existed between animals that produced language and animals that did not arose in large part from our narrow view of the abilities of nonlinguistic animals. Now that we are crediting them with greater mental skills, we can see more clearly how the language we have arises from the platform of thinking and communication that we share with them (or, if you want to cut it more finely, from the many platforms we share, each resting on the other, mammalian arising from reptilian, and so on). The common platform arises from common genes.

Since Darwin's time we have come to understand that not only has all life descended from the same ancestor but many features that arose in more recent ancestors are still shared between us, being built by the same genes. We can see that biologically we are basically African apes who only recently left the motherland. And our most distant human ancestors have been located in time and space. All of us alive today share at least one grandmother who lived 150,000 years ago in East Africa.[4] We also share at least one grandfather, an African man who lived 60,000 years ago.[5]

We see today that differences in complexity between life-forms arise more from the way that genes interact with one another than from their raw number. It's clear that the notion of a genome as a blueprint—so popular only five years ago—is at best inadequate and at worst completely misleading. Instead of following straightforward predetermined plans, genes operate in a dynamic fashion. Many genes respond to the experience of the organism they are building, and they can be switched off or on by other genes or by the effects of the environment.

If a gene comes on in the right cell at the right stage of development, it has a beneficial effect. The same gene acting at the wrong time or in the wrong place can be devastating. Vision, for example, doesn't just unfold automatically in certain animals. The animals need to be exposed to light for the right gene to start building the ability to see. Moreover, different genes have dominion over different body parts. HOX genes divide up the body plan of organisms, with each affecting a certain segment. Some genes are noted for their effect on other genes. These manager genes turn numbers of other genes on and off, and in this way changes in a single gene can cause chain reactions of gene expression.

What we have learned about genes has allowed us to understand that we are not so much things merely existing in the world as beings in constant interaction with the world. If you took this idea to an extreme and imagined that you grew up on another planet, the essentially dynamic nature of animal building by genes *and* environment might mean you'd look very different. Cloned plants that have exactly the same genome can look like very different specimens if planted at different altitudes. In the same way, if you had grown up on a planet with lower gravity or one that was more distant from the sun and had a lower oxygen concentration, you might be incredibly tall, or short, or weedy, or blind . . . or maybe you'd have a supersized brain. If you took your African ape genome and cultured it on yet another planet, maybe the resulting *you* would have translucent skin. The point is that although we experience ourselves in some sense as finished or perfected, we are not in any way *intended*. There is no blueprint for what humans are meant to be. And as this moment is merely one moment in the past and future history of our evolutionary lineage, your life right now is merely an instant in the past and future history of the interaction between your genome and your environment.

At the time the discovery of FOXP2 was announced, Faraneh Vargha-Khadem said that she didn't believe it was accurate to call it a language or grammar gene. As she explained, "The core deficits of the FOXP2 gene have much more to do with speech and articulation than with the more complex aspects of language." Certainly it has turned out to be much more complicated than a single-function grammar gene.

FOXP2 is the kind of gene that turns a tree of other genes on and off, so there is no one-to-one correspondence between it and a single trait. As mentioned earlier, it is also expressed in the heart, lungs, and other tissues.[6] For that reason calling FOXP2 a language gene is a little like calling gravity a force that makes apples fall from trees. It's true enough, but it's hardly the whole story. This fundamental truth about genes, in addition to the way that some genes produce cascading changes in others (as opposed to the purely atomistic "gene + gene + gene = discrete trait" idea), has made it increasingly difficult for skeptics to resist new ideas about language evolution.

One of the most exciting things about the FOXP2 discovery was that it seemed to be more than just a gene that could block normal language development (in the same way that, hypothetically, if your mouth didn't form, you wouldn't be able to speak). It looked, rather, as if it had some role in actually building language. In the ensuing years evidence for this has accumulated as other groups have begun to study the effects of the gene in different animals. Although our version of FOXP2 is unique to us, it is highly conserved between species, and in fact predates the dinosaurs. Though there is no direct relationship between possession of the gene and fully having language, FOXP2 does play a role in the communication of a number of different animals.

Scientists say that in humans and songbirds, the gene is 98 percent the same. FoxP2 (nonhuman versions) appears to play a significant role in the learning and expression of song in birds like the zebra finch; its expression increases in certain brain areas at the developmental stage when the birds are learning how to sing. In addition, the expression of FoxP2 in canaries varies seasonally and correlates with a change in song.

The mouse and human versions of the gene are even more alike than the human and songbird versions, and it's recently been demonstrated that FoxP2 affects the vocalizations of mice.[7] Scientists at Mount Sinai Hospital in New York showed that while mice with only one normal FoxP2 had some general developmental delays, more strikingly their patterns of vocalization were affected.[8] Typically, if a mouse pup is separated from its mother, it will produce cries that are above the range of human hearing. (It was only a few years ago that we learned mice produce sound in the ultrasonic range. In 2005 scientists at Washington University discovered that male mice sing to females in the ultrasonic range.) The purpose of the pup's ultrasonic cries is to alert its mother to its whereabouts. Mice with only one working copy of FoxP2 produced far fewer vocalizations when separated from their mothers than normal mice. FoxP2 seems to play a role in both learned and innate vocal production. (The Mount Sinai researchers found that mice with disrupted versions of both of their FoxP2 genes had severe motor difficulties, lacked crucial vocalizations, and died prematurely.)

Even though language ability is not contained in one or two genes and somehow generated out of them, the FOXP2 work is compelling evidence that we need certain genes to have structured communication—and that human communication, of which language constitutes a huge part, depends in some measure on the same genetic foundations that animal communication does.

Gary Marcus, a professor of psychology at New York University and author of *The Birth of the Mind,* has worked closely with Simon Fisher, one of the geneticists known for FOXP2 research. Marcus explained FOXP2 by way of comparison to another gene, PAX6:

> PAX6 is what is called a master control gene—a gene that achieves great influence by guiding the actions of other genes. Strictly speaking, what PAX6 does is the same sort of thing that any other gene does: it gives a template for building a particular protein, and information about when and where that protein should be built. But the protein that PAX6 governs influences the expression of other genes, telling them when and where other genes do their thing. And because it's atop (or at least close to the top) of a hierarchy, PAX6 can have a huge influence.
>
> One experiment showed that by switching on PAX6 in the right place on a fruit fly's antenna, the fly can grow a whole extra eye in an entirely new place. FOXP2 may or may not be so high up the food chain, but like PAX6 it clearly does modulate other genes; if it's not a CEO, it at least seems to be an important middle-level manager. The broader lesson is that all genes work as parts of hierarchies or cascades. PAX6 isn't "the eye gene." It's a gene that can spawn an eye by influencing thousands of other genes. FOXP2 isn't "the language gene" but it may have a profound influence by regulating the actions of many other genes.

After the discovery of FOXP2's language effects, Steven Pinker hailed the possibilities for a new science: cognitive genetics. Vargha-Khadem and her colleagues called it neurogenetics. Whatever this new field ends up being named, the next century will be an exciting time of determining the closeness of the weave of genes, brains, and behavior. The old nature-versus-nurture debate will finally be shucked off and left behind.

In working out the way genes build linguistic brains, one of this new science's greatest challenges is determining how experience affects the spread of job specialization across the brain. The dynamic interplay between genes and experience as it propels a creature through conception, development, sexual maturity, parenthood, and eventually death is greatly complicated by brain plasticity—which must itself, presumably, be underwritten by genes. Solving the mystery of language and its evolution will involve working out what is innately specified and what alternative routes to processing the same kind of data are enabled by plasticity.

As the field progresses, we will discover more about the reach of FOXP2. In his most recent book, *Toward an Evolutionary Biology of Language,* Philip Lieberman notes that in addition to vocal learning, "humans possess more cognitive flexibility than other species." He argues that FOXP2 also underlies this trait, which itself gives rise to creative thinking, language, voluntary motor control in speech, and, perhaps, dancing.

The different research projects reviewed in this book do not line up perfectly with one another; still, much of this work inhabits the same intellectual space, and together it promises to explain at least some of the larger language evolution story.

When examined as a whole, the studies presented here signal a profound change of mood in the scientific community. In most disciplines the focus used to be on the separateness of animals and humans, that gulf being marked most strikingly by language. But over the last few decades, the emphasis has switched to investigating the continuity of life *in addition to* clarifying the boundaries that lie between species. We no longer have a sense that we are standing apart from all animal life and that language is a discrete, singular ability that isolates us.

Despite the initial controversy connected with examining the mental life of nonhuman animals, once this research began (in every field in which it's been approached), it didn't take scholars long to discover that thinking is a widely spread characteristic of many forms of life. In addition, in many animals there is some lexical ability, a capacity for simple, meaningful structure, elements of culture, and the ability to imitate and learn. In animals closely related to us, the rudimentary beginnings of vocal control are evident. Although language evolution is a relatively new field, it has brought together this research from many disciplines in a completely new way.

Part of the field's struggle is that the very language used to get at these ideas does not serve it well. Language evolution research has illuminated a complicated geometry of species, traits, and relationships, and in the face of this newly defined space words like "uniqueness," "innateness," and "instinct" have come to mean everything and nothing. Those terms are still bandied about to explain the disagreements between people working on language evolution, but in fact everyone agrees there is linguistic innateness, and everyone agrees there is *something* unique about language.

Language has to be partly innate, simply because human babies are born with the ability to learn the language of their parents. While this can justifiably be called a language instinct, there is no one gene compelling us to produce language. Instead, a set of genetic settings gives rise to a set of behaviors and perceptual and cognitive biases, some of which may be more general and others of which are more language-specific.

Language is unique in that there are no other animals with which we converse, no matter what language we are speaking. And yet the miracle of this research has been the realization that what is unique from one perspective may be constructed of mostly old parts from another.

All the work in genetics, neuroscience, ethology, biology, and linguistics has emphasized both the undeniable separateness and the powerful continuity of language. We are not the only animals that live within a world of meaning. And yet no other animal mimics in quite the way we do, no animal gestures like we do, no other animal is able to produce such an ordered flurry of distinct and meaningful bites of sound, and certainly no other animal puts all of this together and communicates it in the same way we do.

In their completely different approaches, by building on the work begun by Noam Chomsky, Sue Savage-Rumbaugh, Steven Pinker and Paul
Bloom, and Philip Lieberman, most researchers described in part 2 have
emphasized the same important fact of evolution—having evolved means
that you are less a creation than an accretion. You are a piled-up assemblage
of systems and organs (some of which work better than others), and because of this, focusing only on sameness or only on difference doesn't take
us very far.

Like biology, language is constituted of an aggregate of different traits
and processes that have developed over time. There was no one moment at
which humans became definably human, just as language did not appear
suddenly from the ether. As important as the shared traits that we use as
the basis of language are, so too are the parts that are different. In the end,
you have to be human to have human language.

Investigating the language suite helps us identify the way these traits
and behaviors are wondrously assembled by evolution into an ability to
learn language. What this research does not explain, though, is how language itself came to exist. Indeed, humans won't speak or produce language unless they are taught to do so, which means that our remarkable
capacity doesn't amount to much at all if someone isn't there to provide a
model for how to use it. In order to understand this conundrum, you have
to look at how the old parts, the shared parts, and the new parts have
wound together in humans to produce this novel ability to learn language.

III. WHAT EVOLVES?

When you consider language for any length of time, you come to realize that for its users, language is the operating system of the world.[1] And after barely two decades of research, it is now undeniable that many of the traits implicated in the learning and use of human language are much older than humanity itself. So how did these particular abilities coalesce to produce language as we have it today?

Evolution is a slow and dirty process, and it's difficult to see how something so complex can arise from something so unpremeditated. But it is only because of the opportunistic zigs and zags that biology took through time that we now have words and rules and their infinite permutations, from tedious political speeches to information-packed instruction manuals, from the irresistible oomph of "WAR! (What is it good for?)" to the dirty word someone once whispered in your ear.

The mechanics of evolution mean that humans became the linguistic species through a purposeful but not perfect process. The purpose was not to create modern language per se, but to provide an advantage in staying alive. Nature selected your father and mother and their parents for survival. It selected their parents before them, their ape parents before them, and their lizard parents before them. The long line of specific individuals that precedes you was not fated at birth to survive and pass on a selection of its genes, but somehow it managed to do so. And here you are—the language-rich result of the haphazard, mostly wordless path they fashioned.

When did language begin? Its foundations can be traced back to our common ancestor with primordial lizards. Yet what we recognize as language today took shape sometime in the last six million years. It's clear just from the distribution of elements of the language suite over different animal species and throughout the lineage of the human species that language did not evolve overnight, turning us from *animals* into *people*. The mere fact that different traits are shared with different animals suggests that language came together in bits and pieces, step by step.

Just by looking at what is shared and what isn't, you can start to glean the outline of a trajectory through time from less linguistic to more linguistic. How old is gesture? At least as old as our shared ancestor with other great apes. How old is simple syntax? Perhaps as old as our common ancestor with monkeys, which lived forty-five million years ago. In addition to the fact that traits have changed, it is useful to examine the different ways they have changed over time. When you look at how these pieces evolved, you can start to narrow down why they might have evolved.

12. *Species evolve*

The beauty of comparing the minds and behaviors of humans with those of other animals is that it illuminates a past so distant that it is almost unimaginable. For a trait like language, which leaves behind no fossils, this method serves us particularly well. Even though our common ancestor with chimpanzees lived as much as six million years ago, the extensive research that has been conducted on chimpanzees today enables us to make useful inferences about traits this creature may have had. Chimpanzees and bonobos don't seem to have changed a lot in the last six million years (certainly not as much as we have), so when Kanzi demonstrates the ability to produce or comprehend language at the level of a young human child, it suggests that humans and their closest relatives have been bequeathed a common set of skills that could be used to produce something like language as we know it.

The same is true of monkeys; our common ancestor with the putty-nosed monkey (which utilizes a simple rule to create new meaning out of two separate sounds) lived about thirty million years ago. Indeed, we can trace a common heritage with all of our monkey cousins—the baboons, with their one-sided social syntax, and the Diana monkeys, which make such savvy use of Campbell's monkey calls. It makes sense to assume that even those many million years ago, some animal evolved the trick of combining sounds—sounds it heard or maybe sounds it produced—to create meaning. It may be that our common ancestor with monkeys had only some very limited form of this ability. Regardless, the many language-related abilities that monkeys possess suggest that language as a whole didn't simply spring intact from the head of *Homo*. Its foundations were around long before we were. Alex the parrot, Lou Herman's dolphins, humpback whales—all these creatures help us retrace the long and winding trail from wordlessness to linguistic meaning.

Yet even though the common mental platform we share with other animals turns out to be both deep and wide, a lot has happened since our ancestors split from the ancestors of modern-day chimpanzees and bonobos. Humans have much greater control over the muscles of the face and mouth, the brain has developed along an unusual trajectory, we use a special mental device called a word, and we are exceptional at tracking structure through time. Many of these accomplishments and their consequences—thinking, speech and complicated syntax, rhyming couplets, corny jokes, self-help manuals—were significantly refined in the last six million years.

Unfortunately, there are no animals alive today with which we have a common ancestor from this time range, which means there are no living creatures more closely related to us than chimpanzees with which to make comparisons. We did once, however, have many closer cousins. The creature that was first cousin to the grandparents of all modern-day chimpanzee and bonobos, the creature from which we eventually descended, spawned a number of different species. Unfortunately, all of these branches of the family have died—some relatively recently.

The only evidence we have for the existence of these closer cousins and their relationship with us comes from the fossil record. For the most part, we have identified these relatives by comparing their skeletal features with ours. We imagine who they were and what their lives were like by measuring traits like the volume of the cranium. Cranial size is a clue to brain size, and given our knowledge about how brains work, brain size means we can make some guesses about neural organization. Changes in leg and pelvic bones, as well as the curvature and orientation of the spine with respect to the skull, indicate whether the bone's owners were partly or fully bipedal. We can develop the picture further by examining the artifacts that accompany their remains, if any. We have techniques for modeling the weather of prehistory, so we can establish whether our ancestors lived in warmth or in cold. We also take into account the animal bones and fossils found from the same time period. Because many of them bear telltale signs of having been hunted and consumed, we know a lot about what our various grandparents ate (or what ate them). It is also possible to detect traces of hearths long burned, which tells us who was using fire on a regular basis.

Most of our understanding of human genetic history has come from comparing the DNA of humans across the globe and tracing it back to common ancestors. Occasionally, we can analyze DNA from nonfossilized bony remains. In at least one case, this has told us whether a humanlike fossil came from a direct ancestor. Altogether, this evidence points to many different human relatives—our common ancestors with these creatures lived five, four, two, and even just a half million years ago. As recently as twenty-eight thousand years ago, there lived creatures that were much closer kin to us than chimpanzees are—so close that if you were standing near one of them on a New York City subway platform you might not look twice.

Some obvious general patterns can be discerned in the family tree over the last half-dozen million years. As you'd expect, the further back you go and the closer you get to our common ancestor with chimps, the more apelike our own human ancestors and their cousins are. One of the best candidates for the most distant human ancestor that is not shared with chimpanzees and bonobos is called *Sahelanthropus tchadensis,* or Toumai. Toumai had widely spaced eyes and a small, chimpanzee-sized brain; its face was flatter than a chimpanzee's and pushed outward more than a human's. Toumai's canine teeth were smaller and more humanlike as well. Toumai had a huge browridge, another primitive hominid characteristic, and it lived in the forest on the edge of Lake Chad. So far, only pieces of Toumai's skull have been found, so it's not clear whether it was bipedal or not.

Another distant ancestor (whether direct or more like a great-aunt we don't yet know) is a six-million-year-old Kenyan species known as *Orrorin tugenensis,* thought to have walked on two legs for at least some of the time. The remains of *Orrorin tugenensis,* like those of most of our more recent ancestors, are fragmentary and far-flung. Only twenty-two traces of the *O. tugenensis* family have ever been found, mostly teeth and pieces of limb bone. Despite how different they seem to us today, these animals were closer to us than modern chimpanzees are.

The further along the branches you go and the nearer you get to modern times, the more recognizable the members of the family tree become. Around four million years ago, our ancestors left the forests and moved out to the savanna, where they remained for a very long time. This emigration marks the birth of the fully walking ape. All other apes, even if they can walk bipedally for short periods, primarily move about on four limbs. In 1978, Mary Leakey, of the famous family of paleoanthropologists, stumbled across a line of footprints in Tanzania that date to 3.6 million years ago. Bipedal apes are called hominids, and it's possible to see from the trail that they left that three hominids once strolled together across wet volcanic ash.

This period is characterized by an extensively populated branch of the australopithecines, including *Australopithecus anamensis, Australopithecus africanus, Australopithecus boisei,* and *Australopithecus afarensis*—the famous Lucy, whose close kin lived from around 3.6 million to 2.9 million years ago. Many of these species dwelled side by side in Africa, where they all remained. Lucy's skull was more chimpanzee-like than human, but her canine teeth, though more pointed than a human's, were much smaller than those of other apes. In size, shape, and relative proportions, her leg and pelvic bones were clearly more human, and she was bipedal. Another important *Australopithecus afarensis* discovery was announced in 2006. In Ethiopia, not too far from where Lucy was found, the remains of a small child, dubbed Dikika ("nipple" in the local language, after a nearby hill), were discovered. Scientists pieced together Dikika's face (with a full set of milk teeth, as well as unerupted adult teeth), a hyoid bone, complete rib cage, some fingers, and parts of her legs, including knees and a foot. She was three when she died 3.3 million years ago—the most complete skeleton of her species ever found. Researchers think that Lucy's and Dikika's family are very distant, direct ancestors of modern humans. (The other australopithecines are cousins to *A. afarensis,* parallel lineages that eventually died out.) Not long after the time that Lucy and Dikika walked the earth, our ancestors and their cousins began to develop tools; the first in the australopithecine-hominid lineage that we know of were simple stone flakes.

Between 2.5 and 1.8 million years ago the first species that we would call *Homo* are detected. Around this time, there were at least four branches of the family—*Homo habilis, Homo rudolfensis, Paranthropus boisei,* and *Homo ergaster. Homo ergaster* is our grandparent; the others are our aunts and uncles. (While *Homo habilis* was not a human ancestor, many *habilis* remains have been found with stone tools. Tool use is therefore a more general family trait than a feature of the specific lineage that produced modern humans.) Some hominids had a brain almost twice the size of the australopithecines'. *H. ergaster* is the first creature in the line from the chimpanzee-human ancestor to have a basically modern human body form—tall and upright. *H. ergaster* was also the first human ancestor to start traveling beyond Africa; its fossilized remains have been found in China and Java, where it lived about 1.8 million years ago.

Turkana Boy is the best-known example of a *Homo ergaster* specimen that looks like a human as we know it. Almost his entire skeleton has been recovered. He lived about 1.6 million years ago, and he was about twelve years old when he died. His brain size was double that of chimpanzees, although it was still significantly smaller than ours. Technological innovation had pretty much remained the same since the flake appeared some 2.5 million years ago, but approximately 1.5 million years ago, not too long after Turkana Boy's death, the hand ax was invented, and it appears from the archaeological record that it was the dominant tool for about a million years.

H. ergaster gave rise to *Homo heidelbergensis,* who invented the prepared-core tool, in which a stone was shaped and then struck once to produce a finished tool. The significance of the prepared-core technique is that it involves using a mental template with which to create many copies of the same tool. Both the Neanderthals and humans descended from *H. heidelbergensis,* though for a long time it was thought that humans descended from Neanderthals. In 1997, however, a team of geneticists announced that they had sequenced the mitochondrial DNA of a Neanderthal bone (mtDNA is passed from mother to child and is used to track female ancestors) and found that, for at least this part of the genome, there was so much difference between us and the Neanderthals that they could not be our direct forebears.

In fact, Neanderthals came from the branch of the family that left Africa long before our more immediate ancestors did. They lived across

Europe and in western Asia for at least 200,000 years. They were excellent
stone workers and survived well in cold, harsh climates. The Neanderthals
buried their dead, at least some of the time. They created stone-tipped
spears and hunted large game, killing animals as big as rhinoceroses. For
most of their time on earth, Neanderthal culture was fairly static; the same
tools were used for thousands and thousands of years. (Imagine if clay
tablet and stylus were the only ways to write that we'd invented for 2,000
years.) Scattered evidence suggests that toward the end of their time, after
living a few hundred thousand years in the same way, they began to use
fire, they possibly made flutelike musical instruments, and they even fashioned ornaments like pendants from the teeth of bears, wolves, and deer.

Our direct ancestors left Africa around sixty thousand years ago and,
after taking thousands of years to reach Europe, coexisted with the Neanderthals there, until the latter died out twenty-eight thousand years ago.
The Neanderthal extinction is generally attributed to either too much
competition (clever, aggressive Homo sapiens outmaneuvered them) or too
much loving (we interbred with them and eventually swamped their
genome with our far larger population).[1] It's also been suggested that in
the same way that Western Europeans caused the decimation of indigenous
populations when they made first contact in the last few centuries, Homo
sapiens may have brought new diseases into the Neanderthal world and thus
contributed to their decline. Most recently, Paul Mellars, a University of
Cambridge archaeologist, and colleagues found evidence that a sudden climate shift, when the temperature dropped as much as 8°C, precipitated
the demise of the species.

As the perceived mental gap between humans and animals has nar-
rowed, so it has for modern Homo sapiens and their recent ancestors. For a
long time Neanderthals were considered brutish, unintelligent creatures
with no language or symbol use. But as the anthropological evidence has
accumulated, they have undergone something of an image upgrade. It's become clear in recent years that even though Neanderthals didn't have as
rich a culture as we did, what they eventually developed before they went
extinct was fairly sophisticated. It's worth keeping in mind that although
the Neanderthals disappeared soon after humans arrived, our species has
yet to prove that it has even half the longevity outside of Africa that they did.

Another cousin, thought to have descended from *Homo ergaster,* was discovered only in 2003. The scientific world was shocked by the news that on the island of Flores in Indonesia a team of scientists had unearthed the remains of a creature they called *Homo floresiensis.* Prior to this no one had suspected that humans had once had relatives as closely related as Neanderthals. Nicknamed the hobbit, *H. floresiensis* is an interesting contrast to the Neanderthal. While the northern branch of the family was large, thickset, and reputed to have had larger brains than ours, the hobbits were mini-hominids, reaching only a meter's height at adulthood. The remains of seven different *H. floresiensis* individuals were found; they died between 95,000 and 13,000 years ago. Flint blades were discovered alongside the remains, indicating that the hobbits were also tool users. Scientists say the *H. floresiensis* tools are stylistically similar to a cache of 800,000-year-old *Homo ergaster* tools that had been found nearby, suggesting they inherited the technique from their *H. ergaster* ancestors.[2] The last of the hobbits disappeared at the same time that a nearby volcano erupted, so it's believed that this catastrophe led to the species' extinction. The individual that died 13,000 years ago was our last, closest cousin.[3]

Homo sapiens emerged as a single small population around 200,000 years ago in Africa, and while there are no examples of nonfunctional symbol use of any kind before this time, sparse but clear examples begin to appear from this point on. In 2003 the skulls of two adults and a child who probably lived about 165,000 years ago were found in Herto, Ethiopia. These humans resemble us in many ways, with only minor differences: they were somewhat larger overall and had a slightly protruding browridge. Most interestingly, their skulls had been flayed and ritually incised after death. It's possible the practice is related to the rites of some modern-day groups that worship their ancestors in this way.

Altogether, archaeological and paleoanthropological evidence suggests that *Homo sapiens* remained relatively stable for around a hundred thousand years. Then, between sixty thousand and eighty thousand years ago, there was a dramatic expansion of certain genetic lineages in the African population. At the same time there was a striking change in technology and culture. *Homo sapiens* do not appear to have changed physically during this period, yet they began to produce many more types of unambiguous symbols. New forms of tools such as those for scraping skin and shaping bone and wood appeared, and novel techniques for flaking stone to make tools can be deduced. Some anthropologists believe that at least some of the sharpened stone and bone tools created at this time could actually have been the tips of arrows, indicating the invention of archery and presumably the ability to access more and better food. In addition, many traces of art, such as perforated shells for jewelry and other kinds of decoration, have been found. The oldest known examples of human art—two pieces of ocher with a hatching design clearly carved into them—were created seventy-seven thousand years ago and recently recovered in South Africa's Blombos Cave. It appears that humans were also engaging in some kind of trade at the time. The shells, for instance, had been clearly transported from other locations to the sites where they were ultimately found.

Some small groups of *H. sapiens* left Africa and settled in places like Israel a hundred thousand years ago, but all of these colonies eventually died out. Our direct fathers and mothers left Africa only sixty thousand years ago, soon after their cultural and technological shift, and they successfully introduced their new way of living everywhere they established a foothold. Everyone alive today descended from this small band of travelers.

A number of possible routes have been suggested for this African exodus. The first modern humans may have departed via North Africa and then split, some going west to Europe and the rest heading east to Asia. The other likely route out of Africa was via Ethiopia and was essentially a coastal route through southern Asia all the way to Australia. This group colonized Europe from western Asia, some fifty to forty thousand years ago.[4] (It's possible that these humans bred with the relatives who'd left earlier, around the 100,000-year mark—or that the initial small groups of colonists simply died out, taking their slightly older genome with them.) Paul Mellars notes that if this coastal route was the first successful exit from Africa, then rising sea levels since then mean that most traces of this journey now lie under as much as one hundred meters of water.[5]

Much is made of the brain's plasticity, but the recent, rapid spread of *H. sapiens* across the globe powerfully illustrates how plastic the body is over generational time. Wherever a trail was blazed and settlers were left behind along the way, the human form shrank or expanded or somehow changed to accommodate whatever harsh environment it found itself in. When we left Africa, we were tall, in contrast to the Chukchi, who live inside the Arctic Circle and who descend from the pioneers of that first exodus. The geneticist Spencer Wells has been rebuilding the recent history of humanity by tracking the Y chromosome across the world and back in time up to sixty thousand years. He visited the Chukchi, and noted that even though the nighttime temperature in their homeland can fall as low as $-70°C$, the people have adapted in many ways. In addition to cultural innovations that enable them to live somewhere so cold, they have changed physically as well. Now they are squat and short-limbed, useful adaptations to the extreme cold in which they live.

As humans spread across the globe, their material and symbolic culture grew richer. By the forty-thousand-year mark, *Homo sapiens* were sculpting from stone, painting in caves, and creating a greater variety of musical instruments and jewelry. They were also ritually burying their dead with grave goods, suggesting that they could imagine a place after death where those items might be useful.

After settlers arrived in Europe forty thousand years ago, it took some thirty thousand years to invent agriculture, and in the ten thousand years that followed (bringing us to the present), agricultural techniques have radiated out across the world. In the last five thousand years we've experimented with architecture, raising edifices that range from the ancient pyramids to the Empire State Building, and in the last three hundred years industrial technology has replaced human labor in countless domains. To be outfitted with a handful of electronic devices is just another day in the life for most Westerners. For modern humans, unlike the Neanderthals or any other species on the planet, culture begets more culture. In the last fifty thousand years, until the present moment, the innovation and replacement of material artifacts have not just accumulated but continually accelerated.

From Toumai through Lucy and the Herto *Homo sapiens* of 165,000 years ago to the small band of humans that left Africa some 60,000 years ago, there are two main theories about the way in which language changed and was elaborated in this bushy, branching, complicated family. As with most evolutionary tales, some scholars see sudden dramatic change from which everything flows, while others are more inclined to detect subtle gradations and tentative steps.

The big genetic bang scenario for culture is most often associated with the archaeologist Richard Klein. In this view, a sudden alteration in the organization of our brains, probably resulting from a genetic mutation, occurred around fifty thousand years ago. This change was the author of all the cultural innovations that followed, as well as the final successful journey from Africa that left humanity spread across the globe. The saltation gave rise to modern language, words and syntax being the cause and the means by which cultural and technological change spread and evolved. Proponents of this theory tend not to consider the cognitive and potentially linguistic capacities of humans from fifty thousand years ago within the larger context of prehuman skills. The implication is more that an *all*—language and culture—sprang from a *nothing*.

If all the significant developments as far as symbol use is concerned took place around fifty thousand years ago, then Neanderthals could not have had symbolic culture. Proponents of this view explain that the evidence of Neanderthals burying their dead, and in at least some cases doing so with grave goods, is accidental. (They only did so to keep the corpses from being eaten by scavengers, and the grave goods were merely swept into the graves by accident.) Examples of the complicated Neanderthal stone tools and jewelry that appeared around forty thousand years ago are generally explained as borrowings from *Homo sapiens* after exposure to their culture and ideas.

Even if you don't subscribe to the notion that a dramatically demarcated revolution occurred, it's clear that a major shift in the history of the human mind was taking place. The most compelling evidence for this "great leap forward" is the proliferation of cultural artifacts that have been discovered from this period. Indeed, symbolic and technological artifacts are often cited as evidence that language existed at some point in the past, but the fact that the largest number of early symbolic artifacts cluster around the fifty-thousand-year mark doesn't mean that humans weren't symbol users before then. As scholars like Terrence Deacon point out, the absence of evidence is not evidence of absence. Deacon cites the case of African Pygmy societies, which leave little more than bone and stone artifacts as traces of their existence, even though their language, cultural traditions, and music are as complicated as those of any iPod-toting modern person.

Deacon takes the contrasting view that symbol use probably began around the two-million-year mark, when our ancestors became bipedal— thus freeing up their hands for tool manufacture and for gesture—and their brains expanded significantly. With that expansion came reorganization of the brain as well, and this, argues Deacon, is a more direct proof of the capacity for symbolic communication than what the archaeological record reveals. (See chapter 14.) Thus symbolic language has been accruing from around the time that the australopithecines were replaced by the hominids, and in the last fifty thousand years it became transformed into its modern incarnation.

Similarly, in the more gradualist view described by the archaeologist Paul Mellars, the traces of art and culture that begin to accumulate in the last 200,000 years are a gradual elaboration of new mental abilities under different pressures (such as dramatic weather changes, and population and social pressures). These pressures did not result from spectacularly anomalous situations, according to Mellars, but were more or less like those that affected later, established agricultural communities.

It's more plausible in this view that Neanderthals did have an indigenous symbolic culture. (How could they not have if their distant ancestors already had the beginnings of one?) At any rate, even if their cultural and technological innovations were borrowings from modern humans, the loans indicate a clear capacity for complex symbol use, if not the inclination to invent it. No other species has borrowed elements of human culture in this way.

Whether symbol use began with hominids or with the australopithecines, it is hard to imagine, given all the language foundations that came together before the 6-million-year split from chimpanzees, that nothing relevant to the development of language took place in the course of the next 5.95 million years, after which modern language suddenly burst forth. Still, we won't know the details until we invent some better methods of answering this question, and until the archaeological record becomes richer. Thanks to some very recent genetic detective work, the period from 200,000 years ago until the present has unexpectedly become much clearer. The most telling finding was announced in 2002—at some point in the last 200,000 years the FOXP2 gene, which has so much to do with vocal communication and learning, changed in humans, and this change was advantageous.

The fossil record, and the ingenuity of the men and women who read it, have yielded many rich stories about the history of the earth and life on it. But the record is shaped by arbitrary forces—when someone perished, and the environment in which they did so (which determines whether their remains become preserved) are all a matter of chance, and the odds against any kind of preservation and fossilization are galactic. In contrast, genetic detective work uses data from modern populations. So when that information is combined with findings from paleoanthropology, archaeology, and psychology, the door of time is thrown wide open.

The most familiar nugget of contemporary genetic knowledge is that we share many genes with other animals. We now know that only a very small number of genes in our genome are specifically human, and we know that genes we share can have mutations that are distinctly human. It is finally clear that genes build individual organisms in really complex ways, but what has happened to our genes through time—before and after the six-million-year mark?

Genomics and population genetics, combined with other traditional sciences, have turned up some decidedly odd findings about gene history. It's thought, for example, that the genomes of humans and other vertebrates contain bacterial genes that were once visited upon their hosts as infections, literally embedding themselves in the genome and permanently altering the genetic building blocks of the host species.[6] Moreover, it was recently announced that a number of living individual humans actually have more genes than others. At least for some of us, the human genome has accumulated genes in generational time. More genes than normal may cause disease, or they may have no effect at all. We know also that some features can devolve as well as evolve. Under certain conditions, flies can be induced to revert to an ancestral state.[7] In the history of their species, wingless stick insects have evolved wings and lost them four separate times. Body hair, according to Richard Dawkins, is one of those traits that may recede or reappear a number of times in the history of a species, as was the case with the mammoths, which rapidly became woolly when the most recent ice ages hit Eurasia.[8] The same is true of the jutting brow of our hominid ancestors. It's one of the features, Dawkins writes, that "hominids seem able to acquire and lose again at the drop of an evolutionary hat."[9]

The primary cause of these changes is genetic mutation. Genes mutate simply as a matter of course, and generally as an artifact of the process of replication. A genetic mutation can have a positive or negative effect on the organism in which it occurs; it can cause a disease (carriers of two copies of the gene that causes sickle hemoglobin cells will be afflicted by sickle-cell anemia), or it can confer greater resistance to disease (carriers of one copy of the sickle-cell gene have greater resistance to malaria), or it can end life.

One of the ways that genetic change spreads through a group of animals is called genetic drift. With drift, mutations that are neither positive nor negative for the individual carrier (in terms of affecting one's ability to produce more offspring) get passed on through the generations. The random drift of negligible genetic changes can eventually spread a mutation through an entire population so that everyone has it. Or a mutation may disappear altogether as its carriers eventually die out.

The other force that affects how a new version of a gene fares and how widely it is passed on is selection. Negative selection removes deleterious genetic variants. In contrast, if a genetic mutation results in a trait that helps its carrier have more offspring (compared with individuals who do not have that genetic mutation), it will spread through a population much more quickly than the casual infiltration of mutations by genetic drift. This is positive selection.

Back in 1990, when Steven Pinker and Paul Bloom championed the investigation of language evolution, they proposed a scenario that could explain the natural selection of language. They based their hypothesis on knowledge from fields like anthropology and psychology, as well as logic, arguing that the complicated design of language, as with the eye, could not have arisen without selection. Recall that they presented this argument in the face of the ubiquitous criticism that because we can never really know if a trait was directly selected or if it arose by accident, let alone why it was selected, throwing out theories to explain adaptation is just an exercise in fiction.

Their endeavor has been vindicated by new statistical techniques that can reveal if a gene was selected or not. Once you know whether a gene has been selected, you can begin to look in more detail at its impact on an organism and substantially narrow down what trait was most likely selected for. The data about genetic changes in general, and about FOXP2 in particular, mark a huge shift in the kind of evidence available for language evolution.

Knowledge of the way that genes work and the ability to determine what's been selected and what has merely drifted have been applied to a comparison of the human and the chimpanzee genome with especially interesting results. These pertain to another important genetic difference between the species: in addition to the different DNA sequence that each has (that famous 2 percent), there are a variety of ways that the same gene can be expressed in the particular organism.[10]

It's clear that in the human lineage, some expression levels have been elevated while others have been significantly reduced. A group of geneticists led by Svante Pääbo (who led the team that sequenced Neanderthal mtDNA and who headed the FOXP2 research) found that the evolutionary change of the expression of genes that shape the heart, liver, and kidneys of humans and chimpanzees is similar, *and* what differences there are in expression evolution have mostly been shaped by negative selection and drift. There is not as much difference between the species in expression in the brain. Said Pääbo, "In the brain, there is a lot of negative selection accounting for the small amount of differences we find, but of the few differences we see, more have occurred on the human lineage than the chimp lineage, suggesting that positive selection may have played a role in human brain evolution."

Another study recently confirmed that many of the differences of gene expression between humans and chimps resulted from much higher levels of expression in genes in the human brain. Like the Pääbo team, this group concluded that changes in many other tissues of the body were random and probably not the result of positive selection.[11] Overall, genetic drift is a much more common process than selection, which makes finding a selected gene especially exciting. It's been estimated that natural selection has had a significant effect on only 9 percent of genes in the human genome.

Because scientists are now able to zoom in on the way a gene version changes and spreads throughout species, the group that discovered FOXP2 started asking questions about how that gene has changed over time. In Leipzig, Wolfgang Enard, who works with Svante Pääbo, presented the history of FOXP2 in the context of the entire human genome.

Showing a slide of President George W. Bush and a chimpanzee, he clarified that there is, in fact, only a 1.2 percent genetic difference between humans and their closest relatives. Between humans and gorillas, there is a 1.7 percent difference, and between us and orangutans there is a 3 percent difference.[12] Moreover, said Enard, most of the differences between us and other animals lie in parts of the genome that are not particularly significant, the junk DNA. Nevertheless, these genomes, which look overwhelmingly similar, produce very different animals: humans have language, and chimpanzees, bonobos, gorillas, and orangutans don't. The differences are not just cognitive, noted Enard. We have AIDS and other apes don't. We have malaria and they don't. We have a doubled maximal life span. We have bipedal walking. And we have a larger and differently proportioned brain.

As for FOXP2, the gene comprises a chain of 715 amino acids. Our common ancestor with mice lived more than seventy million years ago, and our FOXP2 differs from theirs by only three amino acids. Surprisingly, the chimpanzee version of the gene differs from that of mice by only one amino acid, which means that two amino acid changes have occurred in the six million years since humans and chimpanzees split.

The rate of change on FOXP2 is significantly higher in our species than in others. "You rarely get this much change in this amount of time," said Enard. The high rate of turnover suggests that the human form of FOXP2 resulted from positive selection rather than random drift.[13] "It is a rare event to find a selected gene," he said.

Now, crucially, all humans have these two changes, and the age of the bit of the DNA that carries these changes is younger than other parts in the human genome. If this part of our genome is significantly younger and present in all humans today, then it must have spread faster than other parts of the genome. This is equivalent, said Enard, to saying that it must have had an advantage. Enard and colleagues estimated that between fifty thousand to two hundred thousand years ago all living humans had the advantageous version of FOXP2.

Other researchers suspect that FOXP2 is crucial to language evolution partly because of the time frame of its rapid spread through the human population. If FOXP2 mutated to the human version within the last two hundred thousand to fifty thousand years, the mutation coincides perfectly with the acceleration of culture and the migration that spread modern *Homo sapiens* from Africa out across the world, and that meant instead of taking a million years to upgrade our tools, technology now changes every decade or so. Did the mutations, or at least one of them, significantly refine our ability to speak and make complicated syntactic distinctions—resulting in a major change of pace for cultural evolution?

We don't yet know. The correlation between the FOXP2 changes and the blossoming of human culture may be coincidental (or less than direct). As we now know, genes have many different effects. Because the evidence of the KE family shows that FOXP2 is extremely important to language, it's not unreasonable to suspect that the human mutations of FOXP2 were selected for their effect on language. But the changes in the human FOXP2 may have occurred because of their beneficial effect on heart tissue or lung development.

How can the effects that gave the first carriers of the modern FOXP2 such a profound advantage over their peers be identified? In order to take this next step in ancient forensics, Enard and his colleagues will be tinkering with the genome of the mouse. Unlike the team at Mount Sinai that knocked out one working copy of the mouse FoxP2 gene to see how it would affect the animals, Enard and his colleagues are building a mouse with knock-in genes—altering the mouse genome so that it is artificially wound forward along the human line. This involves changing the two amino acid positions from the mouse setting to the human one. What Enard and his colleagues will then do is look at how the changed gene affects gene expression, neuroanatomy, and behavior in the mouse. "What we will not find," said Enard, "is a gene that affects only grammar."

Naturally, it is tempting to see FOXP2 as solely responsible for human language and the last fifty thousand years of rapid cultural rollover. But heralding the new mutations to this ancient gene as some kind of genetic big bang would be tantamount to reviving the old tendency to view language as a singular and discrete "thing" that came about all at once.

Keep in mind that our ancestors used tools for millions of years before this genetic innovation. For these reasons, Gary Marcus advises caution:

> We just have no idea how FOXP2 fits into the space of language-relevant genes as a whole. We know neither which other genes are relevant for language nor how those fit into the oodles of other genes that have been sculpted by positive or negative selection. All told, there are something like thirty-five million base pair differences between human and chimp, and we know some are important for language, some for our physical appearance, others for our immune system, et cetera. We also know some are simply irrelevant. For the most part we simply don't know which are which. I suspect that most of the differences that are essential for language have been subject to strong selective pressures, but the details very much remain a mystery.

Instead, it makes more sense to look at the gene's effects as part of the whole language suite. For Marcus, as for many other scholars, this means treating language as an aggregation of many abilities:

> I think language is probably a patchwork of a dozen or more capacities borrowed from ancestors, ranging from tools for imitation and social understanding to tools for analyzing sounds and sequencing information. Many or even all of those subcomponents probably got further tuned over the course of language evolution, but I would argue language could not have evolved so quickly (or with so little genetic

change) without this sort of broad inherited base. FOXP2 fits nicely with this perspective, since, whatever function it has for humans, it seems to build on a gene that's had something to do with vocal learning for several hundred million years, long before language as such evolved.[14]

As we find more and more evidence of the shared foundations of language, there is much less motivation to search for some crucial single genetic mutation that turned a loose potential for language into language. With this kind of understanding, Steven Pinker made some predictions about research on genes and language in the next decade or two:

> We've found one gene. We've found two other markers for language impairment. We have inheritability studies that suggest that many other forms of language delay—stuttering, dyslexia, and so on—are also inheritable. So let's say in ten years' time, say, ten or fifteen genes have been isolated. Then you apply these statistical tests, and if you have . . . Well, in fact, even if you have one gene that's been the target of selection, that establishes the case, as we already have. But the more genes that have been targeted for selection, the stronger the case would be that language is an adaptation. So I think that's already strong evidence for language being an adaptation, and I predict that's where the debate will eventually be settled.

Ultimately, said Pinker:

> In the case of syntax, it's very unlikely that it's due to a single gene. One of the reasons we know this is that no one has found a case of language impairment where the faculty of language in the narrow sense is completely wiped out in terms of being nullified. We know that at least three genes or genetic markers have been identified for language disorder, and I think most people who work on the genetics of language believe that that's just the tip of the iceberg, and that there are a lot of genes, each with a fairly small effect. That is exactly what you'd

expect of an ability that is polygenic, and that required many evolutionary events to put into place.

One of the most exciting projects in genetic research and human origins was announced in 2006. Svante Pääbo and an American research team
plan to sequence the entire Neanderthal genome. Not only will this provide an incredible framework against which to compare modern-day humans, opening the door even further back in time to our common ancestor
with Neanderthals from about 500,000 years ago, but it may also help
clear up an old but still intense debate. Even among those scholars who believe that Neanderthals may have had some form of language, there is
much disagreement about whether their physiology would have permitted
speech. These arguments generally center on the shape of the Neanderthal
skull, neck, and the remains of the hyoid bone. When the Neanderthal
genome is sequenced, we should be able to look more closely at their version of FOXP2 and see how closely it corresponds to ours.

The synthesis of genetic data with what we know about evolutionary
change gives us one window into the way that genes and traits have accumulated in evolutionary time. But the replication of DNA isn't the only
kind of information exchange in which humans and other animals engage.
Philip Lieberman argues, "Our genetic capacity won't be manifested unless it intersects with culture.

"Look at all the behaviors around now," he says. "Think of what a pilot
has to do to land a commercial airliner. The pilot cannot see the wheels,
but he puts them down nonetheless. Material culture and technologies are
the aggregations of lots of minds. Before the horse was domesticated ten
thousand years ago, people moved at about two to three miles an hour;
then with the steam engine, people started to go thirty miles an hour; now
it's hundreds of miles an hour." Lieberman isn't talking merely about
minds getting together in space to come up with new ideas, but the fact
that today's culture arises from the accumulation of minds through time.
We could never have developed the jet engine in the age of horse travel; we
only got there step by step, propeller by propeller.

13. Culture evolves

Some orangutan groups blow raspberries like goodnight kisses to one another before they bed down on leaf nests that are constructed anew each evening. Dolphin groups off the coast of Australia use sponges to forage. Certain Japanese macaques have invented effective potato-washing techniques that other macaques do not employ. Many chimpanzee groups use tools. Different groups favor different tools—some prefer rock hammers, others wood—as well as different hammering techniques. Chimpanzees will pound their hammers on anvils for up to two hours a day. Some use a fishing technique to get termites with sticks, while chimpanzees in Guinea, western Africa, are the only ones that stand on the top of palm trees and repeatedly beat the center of the tree crown with a branch to make a pulpy soup. In fact, chimpanzees use what amounts to a tool kit. One wild chimpanzee was seen deploying four different implements to get honey from a bee's nest.[1] Animal groups also vary in the way they organize themselves socially. In 2006 a chimpanzee group in Guinea was filmed crossing a road in a highly organized fashion—the alpha males led, acting as crossing guards, while other males brought up the rear so as to get the whole group safely across.[2] It was only fifty years ago that we knew virtually nothing about apes in the wild, let alone dolphins and other animals, but in the years of observation clocked by Jane Goodall and her intellectual descendants, particular animal groups have been shown to have many unique customs.[3]

While the capacity for these preferences is genetically prescribed, the behaviors themselves are not—the chimpanzees that use hammer and anvil are genetically identical to those that do not. As knowledge about behavioral differences between groups of the same species has flourished, scholars have started to regard those differences as essentially cultural.

At its most basic level, culture is merely a group preference for doing things a particular way. As preferences accumulate over time, they become traditions, and these traditions are passed down by a group to its descendants. Just as some human groups prefer spaghetti to rice or high-rise apartments to ranch houses, different animal groups also have different material culture.

The recognition that apes have their own culture has even opened the door to a kind of ape archaeology. In 2002 a group of archaeologists and primatologists announced the discovery of a chimpanzee tool site.[4] Because these apes had a rudimentary material culture, they left evidence of their small-scale civilization from the past. The analysis of the site was the first time that archaeological methods had been applied to a nonhuman culture. In early 2006, members of the same team, along with other colleagues, announced that they had unearthed a 4,300-year-old chimpanzee tool site.[5] The scientists discovered modified stones with food residue attached to them in the Taï National Park, Côte d'Ivoire. The stones predate the settling of human farmers in the area, and suggest that the different varieties of chimpanzee and human tool use originate with our common ancestor (if not before it). The researchers point out that the chimpanzee stone technology is contemporaneous with a local human "Later Stone Age," and therefore indicate a "Chimpanzee Stone Age" (one that apparently continues).

That these animals use tools and develop traditions demonstrates that it is possible for simple culture and technology to arise in the absence of language. Carel van Schaik, who observes orangutans in the Kluet swamp of Borneo, believes that culture and intelligence are inextricably linked. Not only does culture reveal intelligence, he argues, it bootstraps individual animals into greater intelligence. One of the orangutan groups that van Schaik watches is particularly skilled at extracting the rich, nutritious seeds of the Neesia tree. The seeds are encased in a tough pod and protected inside by sharp spikes. Van Schaik's orangutans spend a lot of time inserting twigs into cracks in the husk to release the seeds, and they then tip back their heads and shake the seeds into their mouths.

Only this one group of orangutans employs the Neesia tool technique, and in the Neesia season van Schaik watches them grow fatter by the day.

Other groups have access to the same pods, but their seed retrieval skills are nowhere near as effective. Van Schaik attributes much of his group's success to the fact that they have a particularly high population density, and therefore lots of opportunities for observing the tool use, in addition to which individuals in this group are particularly tolerant of being observed and copied. It's this kind of process, according to van Schaik, that allows animals to stand on the shoulders of previous generations and develop smarter solutions to problems in their world, basically creating new minds out of old brains.[6]

Van Schaik echoes Frans de Waal's wariness about the limitations of lab methods in tapping animal minds: "Our work in the wild shows us that most learning in nature, aside from simple conditioning, may have a social component, at least in primates. In contrast, most laboratory experiments that investigate how animals learn are aimed at revealing the subject's ability for individual learning. Indeed, if the lab psychologist's puzzle were presented under natural conditions, where myriad stimuli compete for attention, the subject might never realize that a problem was waiting to be solved. In the wild, the actions of knowledgeable members of the community serve to focus the attention of the naive animal."

Human culture is an intensely complicated accumulation of techniques and tools. In the same way that an animal's physical development is constrained by its genome, and therefore the genome of its parents, human culture constantly produces new forms of technology and material design by building on what came before. The way we live now is determined not solely by our genes but also by the course of cultural history. Even though the apparent gap between animal and human minds shrinks with each year, there is at this stage little evidence that the social and material traditions of other animals ever move beyond a simple level, in contrast with our own constantly churning culture. Researchers like Simon Kirby at the University of Edinburgh look at the ways in which language is a product of culture as well as biology, asking not just how it evolved but how it might have evolved itself.

Kirby, who completed undergraduate and graduate degrees in the study of language evolution, was appointed lecturer in language evolution at the

University of Edinburgh at thirty-three. This was the first appointment of its kind in the world. Indeed, Kirby is still probably the only academic with language evolution in his job title. Each morning he heads off to his office in the linguistics department, and as he goes through his day, he talks to staff, other lecturers, and students. In lectures, tutorials, and simple hellos in the corridor, Kirby and his interlocutors exchange a certain number of words. If you could zoom out on the department, you would see Kirby and everyone he spoke to zipping around, stopping to connect with one another, and moving off again. Imagine these interactions in fast-forward, the days accelerating into weeks and then years, and all the while see how Kirby and his colleagues talk incessantly. Watch language bubble, build, and evaporate.

Let's assume that as Kirby and his interlocutors get older, they have children, and eventually the children replace them in all the running around and constant talking. Then their children have children. And their children follow in their footsteps. As the talk continues, the language starts to grow and change. Kirby himself may have disappeared relatively early in the process, but the people he spoke to live on, influenced by their conversations with him, and even though they, too, eventually die, the people they spoke to are influenced by them, and indirectly influenced by what Kirby said. Imagine if you could watch this process unfold from the dawn of humanity, watch the first speakers speak and the first listeners listen, and see how meaning and structure develop. Over time, words proliferate and begin to cluster in particular ways, regularities appear, and structural patterns begin to emerge. This grand view of the history of language is a little like what Kirby seeks in his research. His specialty is computer modeling of the evolution of language.

Until the 1990s changes within and between languages could be tracked only by using the comparative method of linguistic reconstruction. But that technique has limitations. No single language from which all the world's dialects are known to have descended has been reconstructed. The comparative method can unearth traces of language from as early as six thousand years ago, but not much further back than that. Computer modeling starts from the opposite end of the language chain. Instead of beginning with contemporary language and reconstructing past versions from it, Kirby creates populations of digital individuals called agents. He hands them some small amount of meaning, maybe a few rules, and then steps back and watches what they do with it.

Jim Hurford, Kirby's supervisor, kicked off the digital modeling of language in the late 1980s. "Jim had read *The Selfish Gene* by Richard Dawkins," said Kirby,

> and in that Dawkins describes a computational model, where these things called biomorphs evolve, you know, bodies and things. Jim read that, and thought, *Wow, I wonder if I could do that for language.* So he started running these simulations on the VAX, an old-fashioned mainframe computer that we had back in the '80s. He would tie up so much of the computing power, the whole department would be paralyzed, and they wouldn't be able to read their e-mail or anything. It was groundbreaking stuff, and he did it really out of a vacuum.
>
> Jim modeled various things, like speech sounds. He built a model about vocabulary and numeral systems, and he did one on the critical period for language learning, which is this idea that we can learn language very easily when we're young, but after a certain age we stop, and our language-learning ability kind of switches off. The question he was trying to understand was: Why on earth did something like that evolve? Why not have the ability to learn language all through your life? And his computational model showed that a critical period did evolve in his agents.

As an undergraduate, Kirby had been deeply inspired by Hurford's lectures. "His ideas about computational modeling really seemed fantastic, and it was just what I wanted to do." So when Kirby finished his undergraduate degree, he enrolled as a Ph.D. student under Hurford.

At around this time, Steven Pinker published *The Language Instinct,* in which he describes Hurford's "critical period" model and refers to Hurford as the world's only computational evolutionary linguist. Since then, the computer modeling of language has boomed. "Every year," said Kirby, "there are more people using the computational approach to language evolution." Today, less than twenty years since Hurford periodically disabled the University of Edinburgh's linguistics department, the school is offering the first degree specifically in the subject, an M.S. on the evolution of language and cognition, and hundreds of researchers are working on computer modeling all over the world.

Even though science has been getting better and better at tracking the elusive clues to our biological language suite, we still don't know how language itself got here in the first place. Computer modeling promises to be a most useful tool in this quest. In addition to the godlike allure of creating populations and then watching them evolve into different kinds of creatures, this technique became so popular so quickly because modeling proposes to answer such questions as: How did the wordlike items that our ancestors used proliferate to become many tens of thousands of words with many rules about how they can be combined today? Why does language have structure, and why does it have its particular structure? How is it that the meaning of a sentence arises from the way it's put together, not just from the meaning of the words alone?

In just a few years computer modeling of language evolution has produced a plethora of findings that are counterintuitive to a traditional view of language. The most fundamental idea driving this research is that there are at least two different kinds of evolution—biological and linguistic, meaning that as we evolved, language evolved on its own path.

Kirby starts his models by building a single individual, and then creating a whole population of them. "I'll have them communicating with each other and transmitting their knowledge culturally over thousands or tens of thousands of generations and very long periods of time. In some extensions of the model, I allow those agents to evolve biologically as well." What he and other researchers in the field have found is that from little things, big things grow. In these accelerated models, from the smallest beginning—agents with the ability to make sound but not words, agents who start out not knowing what other speakers mean—comes incredible structural complexity that looks a lot like language.

This cultural evolution, said Kirby, is simply the repeated learning by individuals of other individuals' behavior:

The idea is that you've got iterated learning whenever your behavior is the result of observing another agent's particular behavior. Language is the perfect example of this. The reason I speak in the way I do is because when I was younger I was around people who spoke and I tried to speak like that. And what we've been finding in our models is, to some extent, that is all you need. It's very surprising. But if you make some very, very simple assumptions like that, you can get linguistic structure to emerge out of nothing—just from the assumption that the agents basically learn to speak on the basis of having seen other populations speak before them.

Strangely enough, the most languagelike structures arise from beginnings that are constrained or not full of information. When Kirby built a model where agents were allowed a lot of exposure to one another's behavior and able to learn all at once pretty much anything they would ever want to say, he found that nothing would actually happen. No linguistic structure emerged from the primordial word soup. In fact, the resultant system of communication looked more like simple animal communication. Kirby discovered that if the agents had only limited access to one another's utterances—either because he made the language so big that they could observe only a small part of it at any one time or because he made sure they listened to only a few sentences at a time—then a lot of syntactic structure would eventually arise over the generations of agents. "It's a kind of irony that you get this complex and structured language precisely when you make it difficult for the agents to learn," he said. "If you make it easy for them, then nothing interesting happens."

It would not be possible for Kirby, or anyone for that matter, to sit down and calculate the ways in which thousands of generations of different individuals may have interacted, and this is what makes digital modeling such a powerful tool. It offers a strong contrast to the armchair models that linguists have used for many years. For example, mainstream linguistics saw language as taking place between an idealized speaker and an idealized hearer. These two were representatives of a population of individuals who spoke pretty much the same language and were basically identical to one another. But this model blurs the distinction between the population and its constituent individuals. Digital modeling allows re-searchers to account for individuals within language communities. Model-ing, then, can consist of at least two tiers of interactions—between individual agents within a population and between populations of these agents.

"If you look at the lifetimes of individuals, you see massive changes in there, from nothing to a full language user," explained Kirby. "It's a hugely complex process that leads from one state to another.[7] Then, on top of that, language changes in a community. So the new thing that's emerging is this desire to link individuals with populations in the model directly, by saying, 'Let's put together lots of agents that are seriously individual, and see what happens when there is a population of these.'"

Because Kirby is working on a vast biological timescale, his models usu-ally involve very simple, idealized aspects of language, like the ordering of words. "They almost seem trivial," he said. Eventually, the models will be-come much more complex, and ideally the particular models that show how language might have evolved from its earliest beginning will mesh with models that show how languages have changed in more recent times—as, for example, how Latin changed into Italian, French, and other Romance languages.

Traditionally linguists have carved up the long history of language into language evolution and more recent language change. Language evolution examined how the human species developed the ability to speak with hu-man language. Language change and growth studies focused on how that first language, once acquired, became thousands of different languages over tens of thousands of years. More and more computer modelers have come to believe that the process is more seamless than that, and language change is to some degree the same as language evolution. The obvious model here is biological life—in the same way that species, once formed, can keep on speciating, the process by which sound and meaning ratchet themselves up into language in the first place leads inevitably to the process by which that language becomes a multitude of languages.

"I would say," Kirby explained, "that the same process or parts of the same process have to be going on. What's tricky about modeling it is the timescales. They are so hugely different. To model biological evolution in a computer you obviously need thousands and thousands of generations, and currently the problem is getting a computer that has the resolution to look at very fine facts about language evolution or language change."

In attempting to incorporate linguistic change in both individuals and populations, Kirby and other modelers like him are actually trying to tease out three different timescales and three different evolutionary processes that contribute to language evolution: two types of linguistic evolution— in the individual and in the population—and biological evolution, tracking how one species becomes another. "That's what is unique about language," said Kirby. "That is what makes it really special in the natural world and probably one of the most complex systems we know of—it's dynamic and adaptive at all three different timescales, the biological, the cultural, and the individual. They are all operating together, and that's where language comes from—out of that interaction."

Kirby and a number of other researchers find one metaphor especially useful for thinking about language: imagine that it is a virus, a nonconscious life-form that evolves independently of the animals infected by it. Just as a standard virus adapts to survival in its physical environment, the language virus adapts to survival in its environment—a complicated landscape that includes the semi-linguistic mind of the infant, the individual mind of the speaking adult, and the collective mind of communicating humans.

According to Terrence Deacon, language and its human host are parasitic upon each other. "Modern humans need the language parasite in order to flourish and reproduce just as much as it needs humans to reproduce."[8] It's an analogy that goes straight to the heart of how much language means to us as a species. If some global disaster killed all humans, there would be no language left. If language suddenly became inaccessible to us, perhaps we would all die, too.

The most exciting implication of the language-as-virus metaphor is the finding that some features of language have less to do with the need of individuals to communicate clearly with one another than with the need of the language virus to ensure its own survival. That is, in the same way that the traits of a particular animal reflect its evolutionary adjustments to survival in a particular environment, so, too, do the features of language structure reflect its struggle to survive in its environment—the human mind. Reproduction is still the driving force of the evolutionary process, but it's not our reproduction: it's the reproduction of language itself.

If language is a virus and its properties are shaped by its drive to survive, then the traditional linguistic goal of reducing all language to a set of rules or parameters is misguided. As Deacon explained, "Languages are more like living organisms than mathematical proofs, so we should study them [in] the way we study organism structure, not as a set of rules."[9] By this light the quirky grammars of the world's languages make about as much sense as a pelican does, and English syntax is as elegant as, say, a panda. You can view any animal purely as a formal system, and you can describe it to a great extent using mathematics, but ultimately living organisms cannot be distilled into rule sets, though each is beautiful, elegant, and perfect in its own way.

If you accept the language-as-virus metaphor, you can't backward-engineer a language-specific mental device simply by looking at the language we have now. If language structure is the result of cultural evolution and accretion, then it's a historical process as well as a mental one. Accordingly, one of Kirby's models showed that a language that has the basic property of compositionality—that is, the meaning of an utterance results from the meaning of its parts and the way they are structured—is going to be more successful at surviving than one that doesn't.[10] Languages that don't develop compositionality are not robust, and they soon die.

"In the model where we don't allow the agents to see all of the language," said Kirby, "structure evolves. The explanation for this is that a structured language can be learned even if you don't see all of it, because you can generalize pieces of it. Whereas an unstructured language, well, you can imagine a big dictionary where every single thing you might ever want to say is listed with a different word. To learn that language, you'd have to see every single word and learn it. But a language that puts words together and allows you to combine them in different ways can be learned from a much smaller set of examples."

As with biological evolution, the road to survival is not straightforward. "What happens," explained Kirby, "if you're forced to learn from a small set of examples is that initially you do very badly, but the language itself adapts in such a way that it is more easily learned by you. We see it happening before our eyes in the simulations. The languages change, and eventually, somewhere along the line, a little pattern will emerge, and that will be learned much more easily than all the other ones. So over time you get this adaptation to the learner by the language. It makes total sense psychologically—the language can't survive if it's not learned."

In 1990 Steven Pinker proposed that our language ability derives from the fact that it is used for communication. Does the virus metaphor completely contradict this approach to language evolution? It doesn't have to. Pinker argued that the appearance of design was evidence of the hand of evolution. This remains relevant for accounts that focus on the survival needs of language. The strong design constraints shown by language in Kirby's model still result from evolution—but the object undergoing that particular evolution is language, not us.[11]

Kirby, Deacon, and the computation modeler Morten Christiansen, a professor at Cornell University in New York State, are especially interested in why language is learned so readily by children. Their approach flips the old notion of poverty of stimulus on its head: if language is driven to survive, and the language learners of the world are children, language must be adapted to the quirks and traits of the child's mind. As Deacon puts it, language is designed to be "particularly infective for the child brain."

So if language in its very structure has all or most of the clues that children require to learn it, then the need for some kind of language organ starts to look dubious. In its strongest version, this approach means there is no support for the argument that grammar is so complicated that children simply can't learn it without a grammar-specific device.

It makes more sense to talk about language learning than about language acquisition, argue Kirby and Christiansen.[12] Their point is simply this: Children do, of course, readily learn language. Instead of beginning with the assumption that this is an impossible task that requires extra explanation, they simply begin by asking, how do they do it?

There is inevitably a human predisposition to language learning. "It's absolutely true that there is an innate component to the process of language learning," said Kirby. "It would be ludicrous to say otherwise. At the most basic level, not every species can speak the languages that we speak, so there must be something there. But in a more subtle sense, we know that we must have some biases. We can't learn everything. There is no such thing as a general-purpose learner, a learner that can be exposed to any task and learn it. So yes, there's linguistic innateness."

The question remains: How much of this bias to learning language is actually language-specific? Said Kirby, "If you added up all of the influence of our learning bias, and all the things that give rise to our learning bias, then the number of things that aren't specific to language but still affect the way we learn language vastly outweigh any language specifics within there." It's more accurate, explain Kirby and Christiansen, to talk of universal bias than of universal grammar.[13]

Another researcher takes up, almost literally, where Jean-Jacques Rousseau left off. Luc Steels heads the Sony Computer Science Laboratories in Paris, which is only a few blocks from the Panthéon, where Rousseau is buried. More than two hundred years after Rousseau wrote about the origin of language, Steels is spearheading a research program that may help us get closer to the answer. He asks: "What are the mental mechanisms and patterns of interaction that you need to get a communication system off the ground?"

Steels's way of imagining the first language users is considerably more practical than his intellectual forebear's. He manages a group of graduate and postdoctoral students, and together they are building creatures—not unlike the inhabitants of Rousseau's primeval forest, the Adam and Eve of language.

In the beginning, Steels's robots had only a single eye and a brain, and their primordial jungle was limited to some basic shapes and colors. Their eyes were black cameras sitting on top of large tripods. Their brains were computers, and their world was a small whiteboard, at which they stared.

Steels made his creatures look at shapes and think about what they saw, and then he encouraged them to talk to one another about it. He is trying to build a linguistic system from the bottom up, as it happened once before, sometime in the last six million years.

Embodiment is crucial. Steels is not modeling language, or a person, or a brain, or a world. His goal is to ground his experiments in hardware that is able to perceive the real physical world. If you go to the lab, you can watch Steels set up his robots and provoke a ricochet of signals between the bodies and the things they perceive; soon a cascade of meaning develops, and a linguistic system emerges before your eyes. Creating a linguistic animal means that, in this context, communication is not a separate, self-contained program, but instead is profoundly shaped by the development of the creature and its world. "These agents are as real as you can get," said Steels. "They are artificial in the sense that they are built by us, but they operate for real in the real world, just like artificial light gives real light with which you can read a book in the dark."

Steels's fundamental motivation is to explore the design of an emerging communication system. "The approach I take," he explained, "is a bit like the Wright brothers, who were trying to understand how flight was possible by building physical aircraft and experimenting with it. They did not try to model birds, nor did they run computer simulations (which would have been difficult at the time . . .). Once you have a theory of aerodynamics, you can take a fresh look at birds and better understand why the wings have a certain shape or why a particular size of bird has the wing span it does." With such insight into the emergence of mental mechanisms underlying a communication system, a dialogue with researchers such as anthropologists, archaeologists, neurobiologists, and historical linguists may contribute ideas to the puzzle of human language evolution.

In most of Steels's "talking heads" experiments, the robots' brains consisted of memory and the ability to produce wordlike sounds. The robots' main way of sensing the world was through vision. Their eyes were directed at simple scenes and objects—a plastic horse, a wooden mannequin—and each robotic individual was forced to find a way to recognize color, segment images, and identify these specific objects. In simpler versions of the experiment the world at which the robots gaze was a whiteboard on which a variety of colored, geometric shapes were fastened. The basic idea is that there is a cycle of back-and-forth between perception of the world and production of language, as the robots adapt and respond to a changing environment in the same way that humans have to.

Steels distributed his robots' bodies throughout the real world, with some going to Paris, London, Tokyo, and Amsterdam, among other cities. The virtual entities occupying the bodies, the agents, were able to teleport through the Internet into specific bodies set up in each lab. Only once they were established inside a body could they communicate about what they saw, and only agents that inhabited the same physical space were allowed to talk to one another. The agents were like strangers at an art gallery, not looking at one another but standing side by side, commenting on the painting before them. This ensured not only that the agents had something to talk about but that they talked about the same physical world.

Steels was inspired by the twentieth-century philosopher Ludwig Wittgenstein's habit of using games to study language. A game captures language in its most basic form, Steels said. It is a simple interaction between individuals within a specific setting. Steels's agents played a guessing game. One agent would pick an object in the world and generate a word for it. Its agent interlocutor had to guess what the word referred to. Each entity took turns at being a speaker or a listener. If one correctly guessed what the other was referring to, the game was successful.

Steels didn't program any word lists or mental and perceptual categories into the agents. They had to segment the images they looked at into sensory data, such as color and position on the board, and then the speaker agent would pick an object based on these data (for example, the red circle in the upper-left part of the board). Then it would choose a word to tell the hearer about the object; that word—for example, "malewina" or "bozopite"—was selected at random. If the listener agent guessed the word's meaning correctly, it might then go on to use it with other robots, and in this way a correspondence between a word and a meaning developed within the population.

Steels found that the game would never get off the ground unless the robots had another channel for communication and verification, so he enabled them to point at the board by moving their camera and zooming in on an area (the other agent could sense the direction the camera was pointing). In one of the largest versions of the talking heads experiment, eight thousand words were generated for five thousand concepts, and a basic vocabulary of fundamental concepts, like up, down, left, right, green, large. There was no central dictionary or record defining each word; they existed only as tokens in the mind of each agent. Meaning was created when agents were able to make perceptually grounded distinctions, such as "left" or "right" and "green" or "red." The distinctions arose when agents identified the object under discussion, separate from other objects in the context.

Since conducting the largest of the talking heads experiments in 1999, Steels and his co-workers have built more complexity into their experiments. One researcher has robots not playing games so much as communicating in order to feel emotion. In another, Steels has robots communicating with ears and vocal tracts to further increase their challenge. The lab is also looking at case marking, tense, open-ended semantics, language processing, and the different types of grammars that can emerge.

Steels is also interested in the way that structure spontaneously arises in biological systems where random behavior is reinforced by positive feedback. He was particularly inspired by Jean-Louis Deneubourg's work on ants. Hundreds, sometimes thousands, of ants organize themselves into long chains when they are carrying material from a food source to their nest. The chains are adaptive: you can sweep away part of one, put objects in its way, remove individual ants or add new ones, and the chain will emerge again until the food source is depleted. There is no central coordinator instructing the ants on what to do and how to organize themselves in the face of disruption. Nevertheless, a greater intelligence—a design—emerges out of the local behavior of many relatively unintelligent individuals. Other systems where order emerges spontaneously from chaos are termite nest building, the growth of cell tissue, the way that cellular slime amoeba form an aggregate entity, and flocking in birds.[14]

The language that evolved in the guessing game has many of the same features as these systems, said Steels. It exhibited an absence of central planning, an adaptation to changing circumstances, and a resilience to the unexpected appearance and disappearance of elements (whether objects or individuals). Meaning and linguistic structure simply arose out of interaction between bodies in space.

Steels has recently taken embodiment to more complicated levels. In 2001 he started work with AIBO robots, which are among the most complex robots ever built.[15] Each AIBO is an independent entity. Steels and his co-workers place the robots in various situations—on a floor with objects like boxes and colored balls, for example—and like the talking heads they must build both a conceptual system and a way of talking about it. The robots develop speaking and hearing processes while constantly trying to map their world (as they move about in it). They also have to work out where another is in space, and if one asks, "Where are you?" and the other answers, "To the left of the box," the first AIBO has to decipher what "left" might mean. His group has also just finished a series of experiments in Tokyo with the QRIO humanoid robot. Working with the QRIO allowed them to implement many of the mechanisms humans use for joint attention, like pointing with a finger.

Because the robots engage in real image analysis (as opposed to being fitted with programs that dictate how to see the world), many errors arise in their interactions. But that's the point, explained Steels. When successful communication does evolve, it shows how language is possible in difficult circumstances. "There is no reason," Steels said, "to think that language processing is any less complicated than vision processing—which is very complicated." He added: "The complexity of language is incredible, but we shouldn't be afraid of that."

As they grope their way through the world, Steels's robots end up evolving rudimentary grammar as well as words and concepts. Syntax arises mainly from a situation of ambiguity. In phrases such as "red ball next to green box," it is not clear to agents whether "red" goes with "box" or "ball" (unless they already have grammar). When an ambiguity like this is detected, the agent will invent a grammatical pattern to make his intended meaning clear to the listener. This suggests to Steels that human language ability is an emergent adaptive system that is created by a basic cognitive mechanism rather than by a genetically endowed language module.

Neither robotic nor digital linguistic systems can tell us exactly how language evolved. Indeed, the communication systems that arise in Kirby's modeling or Steels's experiments may or may not have the characteristics of human languages. What each can do is show how language might have evolved, and this is invaluable data. We can't think these concepts through with our brain alone—instead we had to achieve this stage of technological innovation with computers fast enough to model such complicated processes and robots that can enact them. Kirby's virtual linguistic creatures and Steels's real ones suggest that in order to get to something that looks a lot like language, you may not need a language-specific mental device. Humans do a lot more with language than simple pointing and referring, but in order for language to become established, the ability to perform these steps is essential.

The most elusive part of the language evolution mystery is working out why all these things happened. Why did our species evolve in the way it did? Why does culture evolve the way it does? And even more complicated, how and why do they evolve together? The rollover of language change is thousands of times more rapid than biological evolution. We might find it difficult to talk with English speakers from a thousand years ago, but we wouldn't have any trouble procreating with them. The final and greatest challenge for language evolution is discovering how the language suite and language itself evolved together.

14. Why things evolve

G enes mutate as a matter of course. If the carrier of a mutated gene is lucky, some effect of the new version will improve its chances of having offspring that survive, and then those offspring will have their own successful offspring, and so on and so forth. Every animal alive today stands at the end of a long line of lucky entities that begat lucky entities that begat lucky entities. They may not have been happy or fulfilled or at peace with their lives, but that's not the point.

For a long time people have wondered why a particular trait has evolved. What was it about that trait and the environment in which it arose that meant it was a good thing to have? These considerations have been the most contentious part of the language evolution debate: Why did language evolve?

Part of the problem with posing this question in decades past was that even though scientists were using the same words, they were asking a fundamentally different question. At that time, language was still generally thought of as a single entity. Regarded as such, it left the question truly unanswerable, for different components of language have evolved in different stages in the history of life. If you ask, "Why did the whole thing evolve?" the implication is that it happened all at once, and no evolutionary pressure is up to the task of bringing forth everything from nothing.

The other problem with asking this question is that to some extent you have to imagine the answer. No one can ever know all the details of what happened when our distant ancestors began to talk. The only way to be completely sure is to travel back in time to witness the process, and we can't do that. And there's the problem of language fossils. There are none, at least none as definitive as the femur that Lucy left behind. As Chomsky has pointed out: "There is a rich record of the unhappy fate of highly plausible stories about what might have happened, once something was learned about what did happen—and in cases where far more is understood [than with language evolution]."[1]

However, the same objections could be raised about any attempt to explain the origins of the universe. In *Fire in the Mind,* George Johnson reminds us that the big bang scenario is still only a theory. Nevertheless, the intense layering of evidence and theoretical modifications that have accumulated since it was first proposed have given the theory the heft of unassailable truth. Today, says Johnson, the theory remains a work in progress that underpins the productive work of thousands of astronomers and physicists all over the world.

Cautions against employing "just-so" stories and fairy tales to trace language evolution had great resonance when less data were available about what happened and when it happened in the development of language in evolutionary time. Now the accumulation of evidence from genetics, comparative biology, behavioral studies, linguistics, and neuroscience makes such stories more feasible by placing powerful constraints on them.

With the information scientists now have about gesture, thought, and behavior both in humans and in close and distant species, they are better equipped to carve out the problem space and define the outlines of their story. They know more about where to look for clues and what paths not to take in a possible reconstruction of language evolution. It will never be possible to recover and rebuild every step of the way. But significant steps, major biological traits, and evolutionary landmarks can be identified. And while there are a number of ways in which the facts about humans and life and language evolution can be mapped onto the known evolutionary path that brought us to where we are today, data gathered over the next few years will further refine those conjectures.

In this context the prohibition against asking "why?" is starting to look as unscientific as the kind of fairy tale it once warned against. Indeed, there's something a little disingenuous about the insistence that because you can't prove it, you shouldn't imagine it. Imagination is at the core of the scientific process. All the tests and experiments in the world mean nothing without the hunch or the story—the hypothesis—that kicks the process off. Now, instead of not venturing into the imagination or simply not declaring what they suspect, many scientists in the field of language evolution choose to propose a story *and* be up-front about how much their theory has been informed by data and how much is not yet verifiable.

Michael Arbib, one of the researchers who has investigated mirror neurons, has an idea about what he thinks might have happened and why it might have happened, based on the rigorous work he has carried out on the brain. Arbib's approach is the opposite of the traditional Chomskyan one. Instead of emphasizing the fundamental sameness of language in a search for universals, he is interested in the different ways that people solve problems with language. As he explained: "Once you get beyond the fact that you've got to have words for actions, you've got to have words for objects and the agents that act upon them then, I think, you get into the realm of what people have learned over the centuries to do, rather than something that must be in the brain. People advertising universal grammar focus on what is common. I'm just struck by how varied the approaches people in different communities have to solving communicative problems." Instead of tracing the parameters of language back to genes, Arbib thinks that most of grammar and the way that structure relates to meaning are products of culture. "My feeling is that most of it is probably a tribute to human ingenuity. I mean, kids can surf the Web, and nobody says there's a Web-surfing gene."

Like Lieberman and others, Arbib disputes the idea that language is one big package, a kind of all-or-nothing proposition. When it is conceived of in this way, he said, "I think you make some very foolish claims." The alternative is to take the historical point of view. "You can imagine the first protolanguage as ten words or a hundred words. Then a lot of things can occur over the generations and crystallize out. Language becomes very mysterious if you have to make it a single biological evolutionary leap."

One of Arbib's most important points is that language is not inevitable. He encourages thinking about possible stages by stepping away for a while from the end state—the current form of language. We have it today not because we took one crucial turn at some point in the past but because we took hundreds of crucial turns. And for each of these turns, you can't know that you are going to get language at the end of it. Each step is critical for the value it adds at that point in time. Linguistic evolution was a tumultuous natural experiment that started with a particular brain structure and hundreds of variables—a couple of ice ages, constantly evolving predators and prey, a changing social structure. The process lasted many millennia, there was no control group, there may have been false starts along the way, and the completely unpredictable result of this random experiment was modern language.

The mirror neurons discovery set Arbib on a course that has most recently ended with his fully articulating an idea that many researchers assume but few have examined in detail; that is, language evolution had to occur in a layering of stages. Arbib calls it an ascending spiral. So far he has proposed about ten different stages, though he warns that even a ten-stage theory is still a long way from accounting for all the steps along the journey to language.

Initially, he says, our ancestors must have developed a capacity for complex imitation that went beyond that of even modern apes, and greatly increased the possibility of social transmission of novel skills. Beyond that there must have been some kind of gestural protosign that broke through the fixed set of primate vocalizations and was supported by the mirror system. Gesture, in his view, was an ancient scaffolding on which language started to build. You had to use protosign to build the scaffolding, and then sounds became parasitic. Speech did not arrive directly, and the first gestural steps of language would have been quite simple. "It doesn't make sense to have a full sign language and then go to vocalization," said Arbib. "It's hard to build up a rich tradition just through gesture. You need sound to flesh out that scaffolding." So there were oral and facial gestures as well, maybe some association between lip movements and what lips are often used for, like eating.

Pantomime probably provided the crucial bridge from imitation of practical skills to imitation of the skills required for proto-sign (and much later for language). "The claim is something like this," he explained:

You've got a system for primate calls, but it's closed. You can't add a new call to it. So you use a different system, you go through a different route, to be able to create new patterns of sound that can be paired with new meanings. And then we eventually get to the stage where you can get those sounds and meanings together to create new meanings on the fly. But there must have been an intermediate time when you didn't create new meanings like that. My argument is that if you look at the ability of the hands to move skillfully, then you can imagine that there was an evolutionary advantage in being able to imitate patterns of hand movement, and having imitated patterns of hand movement (once you had a brain in place that could do that), it's a plausible step to begin to use patterns of hand movement for communication—pantomime. And the beauty of pantomime is that if you pantomime carefully, and maybe do it three times when the person doesn't get it, you can convey novel meanings.

Why we didn't ultimately become a species that is constantly engaged in pantomime with no speech, said Arbib, is because "people aren't very good at recognizing someone else's pantomime." As he explained, "It doesn't have to be that you suddenly have a society in which everybody was doing pantomime and conveyed thousands of meanings, but maybe in a particular year, two or three pantomimes were added to the tribe's vocabulary by becoming somewhat ritualized to make them easier to perform and understand."

After this, Arbib suspects that humans developed protospeech:

The story goes (if it went anything like that) that in the end you can't disambiguate pantomime by just doing better pantomime. If I flap my hands to imitate the flapping wings of a bird, do I mean "fly"? Do I mean "bird"? Do I mean "bird flying"? So maybe some genius comes along and invents some way of saying, "Well, if I do this sound and I'm flapping my hands, I mean the bird. If I do another sound while I'm flapping my hands, I mean the flying. You need to make distinctions. So the notion is you got to the stage where a sequence of gestures can convey meaning, and you got across the idea that meaningless gestures are part of conveying meaning. It's no longer pantomime.

Vocalization was involved all along, said Arbib. There may have been stages where the pantomime was entirely vocal. "My purely fictional example is that you bite the piece of fruit, it's sour, you go—" He puckered and made a sucking sound, before continuing:

> The act of genius there is to go from having that as a reaction when it's too late and you've already bitten the fruit, to making that noise, before somebody bites the fruit, to warn them, "Don't waste that fruit; it's too sour to eat."
>
> So the notion is that the pantomime would give you the possibility of conveying a rich sense of meanings. The arbitrary gestures would come in to begin to allow you to save effort and avoid ambiguity. The gesture in the end is conventionalized. It doesn't have to be a fresh pantomime all the time. Then the sound can come into play, and it can begin to become part of an integrated performance. Beyond this, you begin to find certain conventional distinctions that are easier to convey, and then you begin to build a phonology, and then, as you begin to build a phonology, you begin to put those meaningless phonological gestures together to take over more of the conventionalized meaning. I want to claim that this skill was parasitic on increasing manual dexterity and the mirror system that supported it, which increases the cortical motor representation, and then we can expand that to the new use in the vocal system. So I prefer that story at the moment.

Arbib's account could be further elaborated by explaining why the pantomimes were taken up and spread throughout the group. Perhaps it was a case of sexual selection, as Pinker and Bloom suggested in 1990. In this scenario, the mime is a male who impresses females with his linguistic skills, thus creating more opportunity to procreate, having more children, and spreading the predisposition for expression. The same principle explains why male peacocks develop such spectacular tails.

Tecumseh Fitch, on the other hand, argues that sexual selection is particularly unlikely as an explanation for linguistic evolution. As with the peacock, this kind of selection generally results in a marked difference between the sexes with regard to a particular trait. However, not only do men and women both use language, but young females are more adept with language than are young males. Other pressures may have come into play. Perhaps a change in the available game required better hunting techniques, which in turn required more precise language. Maybe the step from one form of protolanguage to another occurred when hominids reached a critical level of population density—just like the orangutans with their Neesia-splitting techniques.

In an interview Chomsky suggested there had to be a point in time when a rewiring of the human brain that allowed people to use recursion took place. Perhaps sixty to seventy thousand years ago in a small hominid group in East Africa, a single individual was born with a genetic mutation. This mutation would have caused a restructuring of the brain and instantly bequeathed the affected person with the capacity for unbounded thought. Linguistic communication would not have begun at this moment, because the individual with the mutation was the only one with the capacity for it. But even a slight advantage spreads quickly throughout a population, and after this new rewiring was passed on to his or her offspring, the entire group would eventually become language-ready.

Is it possible that even though some of these accounts appear mutually exclusive, the researchers are actually describing different stages of a cohesive evolutionary narrative? Yes. It's likely that different parts from many theories will survive in a grand synthesis because within this vast time frame, numerous evolutionary pressures had some effect. Given the way the recent accumulation of data about how the brain works and genetic influences on language have forced researchers to constrain their theories accordingly, a more widespread agreement isn't out of the question in the near future.

At any rate, the question of which specific evolutionary pressures were in play at which moment in time is a less prominent consideration in the field. There is less concern about why language came to be because there is so much to say about *what* came to be and *how* it came to be—which gene changed, which behavior is ancient, and which ability is new? At this point, we must be content to survey all possible answers and acknowledge that in the last six million years many of them probably played a role. The stories can be especially helpful as spurs to testable hypotheses.

In addition to examining the specific pressures, incidents, and abilities that have contributed to the story of language evolution, it is also important to look at co-evolution: the way that human language and the human genome have shaped each other. Co-evolution is the least explored aspect of the mystery. For all the difficulty and challenge of tracing language evolution, working out how species and language arise over time and then provide feedback to each other is probably the hardest part.

Terrence Deacon has grappled with the issue of co-evolution, focusing on the back-and-forth between language and the brain. Recall that he proposed that the beginnings of language and symbol use can be found in the shift from australopithecines to hominids some two million years ago. What preceded this evolutionary shift was the use of flaked stone tools. Deacon argues that it was this tool use that spurred the evolution from one kind of primate to another and, in doing so, created an animal with a predisposition for even more symbol use.

This kind of change, which is called Baldwinian evolution, occurs when the behavior of an animal actually contributes to the environment in which its genetic evolution is shaped. Lactose tolerance is an example of Baldwinian evolution in humans. The ability to digest dairy products in adulthood is most common in groups of people who have been herding animals the longest. In this case, it's a behavior—herding—not a climatic change or some other kind of environmental shift, that contributed to the selection pressures in which a predisposition for lactose tolerance improved reproductive success.

The australopithecine tool use helped to create a world where it was more and more useful to have the genetic predisposition underlying that behavior. The better an individual was at it naturally, the more likely he or she was to survive and have offspring, probably passing this trait on to them, and the more significant that behavior became in the world of the species. It wasn't that our brains got bigger as a result of bipedalism or dietary changes or any other reason, thereby making us clever enough to invent stone tools; rather, we started to use stone tools that are slightly more complicated than the tools chimpanzees use even today, and as a result our brains got bigger.

The co-evolutionary story that began at this time and that continues to this day is one in which the Baldwinian interaction between culture and biology played a particularly significant role. Deacon points out that our brains did not get bigger in the australopithecine-hominid transition in the same way that the surface of an inflating balloon gets bigger all over. It was the forebrain, particularly the cerebellum and the cerebral cortex, that ballooned, while the rest of the brain followed the growth rate seen in other primate brains.[2] In order to unwind the ways that language and the brain have co-evolved, you have to look at the parts of the brain that got bigger, says Deacon, and you have to look at how they got bigger.

The prefrontal cortex corresponds to a small section of the developing brain in the human fetus. When the brain is nothing more than a neural tube, the part of the tube that later turns into the forebrain breaks out of the growth patterns that constrain the rest of the brain. This stretch of tube is controlled by the Otx and Emx genes. The developmental clock that signals to every part of the brain and body when to stop growing has been extended for the regions controlled by these genes. The significance of this altered growth pattern, according to Deacon, is not that human brains are faster and better computers; it means that the balance has been shifted in terms of *what kind of thinking* goes on in the brain. As a result, our learning skills are biased toward certain types of processing and not others.

Acquiring and deploying the particular kinds of connections and structural patterns that characterize language, says Deacon, pose some very unusual learning problems, and the kinds of learning processes that most mammal brains are specialized for are not well equipped to deal with these problems. However, the neural machine that results from the human combination of body and brain growth patterns is one that rather brilliantly performs the computations that underlie language learning.

The fact that language arises from dynamically interacting brain regions with their vastly different evolutionary histories (the more primitive and unchanged along with the parts that have shifted more recently) is another reason why we should not think of language, or even other mental abilities, such as mathematics, as monolithic things. Instead, argues Deacon, they arise out of a "delicate balance of many complementary and competing learning, perceiving and behavioral biases."[3]

Upending these assumptions about brain evolution leads us to a startling conclusion, says Deacon. One of the reasons we haven't been able to work out how language and the brain co-evolved is because we have been asking the wrong question all along. From the beginning, researchers investigating the brain and language have assumed that the brain came first. The usual line of reasoning holds that the brain was selected for increased general intelligence and then it evolved language, which relies on that optimized intelligence. Actually, says Deacon, we should be looking at the effect of language on the brain, as well as the effect of the brain on language.

Generally, the amount of brain tissue devoted to particular types of processing is proportional to the amount of information being processed. Brain regions that serve seeing, smelling, and touching, for example, are matched sizewise to the amount of information that is filtered through our bodies in these senses. One of the crucial differences between the human brain and other mammal brains is that ours is larger overall relative to the body. This leaves a considerable proportion of the human brain, says Deacon, that is not processing information from the outside world in the way that the visual and auditory cortices are.

Even though they are not directly processing sights, sounds, and other senses, the unusually expanded prefrontal brain regions look as if they have been "deluged with some massive new set of . . . inputs."[4] The larger brain region, says Deacon, is "an evolutionary response to a sort of virtual input with increased processing demands."[5] That input, of course, is language.

In this view, language cannot ultimately be treated as a straightforward example of the capabilities of the brain, and we should not be asking "How did the brain evolve language?" Rather, we should ask, "How did language evolve the brain?" Language is the author of itself, says Deacon, and the brain is the smoking gun for language.

The result of the co-evolution of the human brain and language is that we now have an overall cognitive bias toward the "strange associative relationships of language." In this sense our *whole* brain is shaped by language, and many of our cognitive processes are linguistic. What this means, according to Deacon, is that once we have adapted to language, we can't not be language-creatures. For us, everything is symbolic.

Indeed, Deacon explains, the virtual world that we inhabit is as real, sometimes more real, than the physical world. Even the tendency to infer the hand of a designer when faced with complex design (whether it is a deity that has designed all of creation or a special language organ that generates human languages) arises from the fact that we are a symbolic species. Ironically, what makes it hard to discern how language evolved is a result of language having evolved. The worldwide web of words and rules that we inhabit is so vast, contracted, and dense, it's hard to look in from the outside.

Arbib and Deacon seek to illuminate moments in the last six million (and more) years of human evolution. By comparison, the last ten thousand years is a blip. Nevertheless, it is an interesting period in the history of the human brain and also of language. There is some suggestion that our brains may have changed as little as ten thousand years ago, and in fact, become smaller. It will be some time before data on the trajectory of brain growth at this time are solid enough for us to be confident of this change. Generally, it is thought that within the last ten thousand years there has been no obvious anatomical change arising from the drift and selection of genes in our species. The same goes for language. Most language change in this time frame is associated not with obvious biological change in humans but with the movement of human populations and transformation of their lifestyles.

Jared Diamond and Peter Bellwood examined the effect on farming, which independently arose in human communities at least nine different times between 8500 and 2500 B.C.[6] The researchers demonstrate that the advantages of farming over hunter-gatherer lifestyles, including greater access to food, denser populations, and greater resistance to disease, spurred the spread of farming communities, and their culture and language with them. They propose essentially that prehistoric language and genes spread with prehistoric farming, and that tracking one will illuminate the ancient paths taken by the other.

There are many different types of clues to the prehistory of language, and their intersecting relationships are complicated.[7] Here the researcher interested in connecting the long and short arcs traced by language in time must master at least genetic, archaeological, paleoanthropological, linguistic, and geographic evidence. As more researchers engage with this multidimensional problem, we will see ever more clearly how a mental bias gave rise to a language, which became languages, and then rich and sprawling language families.

IV. WHERE NEXT?

15. The future of the debate

Pinker and Bloom's 1990 paper caused a sea change in the attitude toward language evolution, and the early years of research that followed were a time of great exhilaration and puzzlement. The 1996 Evolution of Language conference, organized by Jim Hurford and Chris Knight, was the first in what became a series of biennial meetings for scholars from various disciplines and countries to come together to address this issue. The participants brought their biases and jargon with them, and there was less shared language and understanding than had been hoped. In the end, no synthesis was reached that would get everyone on the same page. In this early period, a great deal of energy was expended in simply justifying the research. As the years went by and more data and ideas accrued in the biennial conferences, and as other conferences also started up, certain questions and methods—those reviewed in parts 2 and 3—emerged as central.

Neither Pinker nor Chomsky said much on the topic in this period. In 2002, however, Chomsky appeared in a panel discussion at the Harvard Evolution of Language conference. Tecumseh Fitch was one of the conference organizers, and Marc Hauser sat on the stage next to Chomsky. Pinker was in the audience, although he, like Chomsky, had not attended other conference presentations. Chomsky suggested that language evolved separately from speech, because deaf children are still able to learn sign language, and he proposed that people use language more for talking to themselves than to talk with other people.

Later that year, Hauser, Chomsky, and Fitch published a paper in *Science* called "The Faculty of Language: What Is It, Who Has It, and How Did It Evolve?" The point of the paper was to provide a framework for fruitful discussion and clear up confusion in the field. It argued that a lot of research vital to an understanding of language and linguistic evolution was typically

ignored or dismissed by linguists, and it also advocated collaboration between researchers from different disciplines.

In an accompanying editorial, titled "Noam's Ark," linguists Thomas
Bever and Mario Montalbetti wrote: "Language is naturally viewed as a
unique feature of being human. Accordingly, the study of what language
is—linguistics—has been very influential, primarily in the social and behavioral sciences . . . Hauser, Chomsky, and Fitch expand the scope of
language study with their demonstration that complex behaviors in animals and non-linguistic behaviors in humans can inform our understanding
of language evolution."

The article inspired many impassioned responses, some as enthusiastic
as Bever and Montalbetti's. Others expressed shock and even rage. "The
Faculty of Language: What Is It, Who Has It, and How Did It Evolve?" gave
the impression, at least to some, that Chomsky had abandoned his old view
of language and swapped sides in the great debate. Derek Bickerton, a
longtime Chomskyan linguist, wrote:

> Into the middle of this confused and confusing situation there appeared
> in the journal *Science* a paper . . . aimed at setting the scientific com-
> munity straight with regard to language evolution. Its magisterial tone
> was surprising, considering how little work any of its authors had
> previously produced in the field, but no more surprising than the
> collaborators themselves: since Hauser was known as a strong con-
> tinuist and Chomsky as a strong discontinuist, it was almost as if Ariel
> Sharon and Yasser Arafat had coauthored a position paper on the Mid-
> dle East. In this paper, practically every aspect of the language faculty
> is treated as pre-existing the emergence of language, except for "nar-
> row syntax" (whether this is the same as, or different from, the old
> "core syntax," we are nowhere told), which consists solely of recur-
> sion. Even recursion is supposed to derive from some prior computa-
> tional mechanism employed by antecedent species for navigation,
> social cognition or some other purpose as yet undetermined, and then
> exapted for syntax; researchers are adjured to start searching for such
> mechanisms.[1]

The reaction to "The Faculty of Language" served as a catalyst in the same way the Pinker and Bloom paper did twelve years earlier. The perception of an allegiance to Chomsky was a lightning rod, although it meant different things to different people. There were two main camps of disagreement. Some critics thought the paper consisted of the same Chomskyan ideas of the last four decades, dressed up as something novel with animal data attached. Taking the completely opposite view, others were angered by what they saw as a retraction of ideas that Chomsky had spent years developing. Depending on their field, researchers suspected either that Chomsky had influenced Hauser and Fitch or that Hauser and Fitch had hijacked Chomsky.

In their *Science* paper, Hauser, Chomsky, and Fitch proposed a two-part model of language, based on a broad faculty of language and a narrow faculty. The broad faculty comprises the narrow faculty, in combination with two other systems. The first consists of the nerves, muscles, and organs that enable us to see, hear, and touch the world around us; it also includes the physical characteristics we use to create and interpret speech, such as the agility of our tongue, the position of our larynx, and our ability to interpret stress and pitch. The second system consists of a creature's knowledge of the world and its capacity to use that knowledge to form intentions and act upon them. The authors called them the sensory-motor and the conceptual-intentional systems.

At a minimum, wrote the authors, the narrow faculty is a computational system that "includes the capacity of recursion." Elsewhere, they described the key component of the narrow faculty as a recursive computational system that generates linguistic structure and maps it onto the two other systems. In this sense, the narrow faculty of language is an interface between recursive computational abilities, the body, and thought.

The authors then presented a distillation of opinion in the field in the form of three distinct hypotheses, using their terminology of a broad and a narrow faculty. In one hypothesis, all components of the broad faculty of language have homologs in other animals, so there is nothing in language that is unique to humans. In an alternate theory, the broad faculty is a uniquely human adaptation. So even if other animals have traits that appear similar to human traits used for language, such as social intelligence or toolmaking, they have been significantly refined in the human lineage and should be considered novel features, specific to humans.

In a third hypothesis of their own, the authors proposed that most of the broad faculty of language is shared with other species, and that any differences in the human and animal traits are quantitative rather than qualitative. They cited experiments conducted by themselves and others showing that animals understand the world in complicated ways. For instance, some birds use the sky and landmarks to help them navigate complex paths; other animals, such as monkeys, recognize and can use in varying degrees abstract ideas like color, number, and geometric relationships; many different species can use mirrors to locate objects, and chimps, bonobos, and orangutans even appear to recognize their own reflections; also, chimpanzees seem to infer from a person's or a fellow chimp's actions what that creature is thinking.

In contrast, the narrow faculty of language is a recent, uniquely human innovation. Hauser, Chomsky, and Fitch noted that even though the recursive mechanisms that underlie syntax may be unique to humans, they are not necessarily unique to language. Instead, this system could be a spandrel, having evolved for something other than communication and still used in nonlinguistic domains. Where did this capacity come from? Perhaps, they wrote, it was initially used for navigating social relationships and only later co-opted by language. They pointed out that because chimpanzees have highly complicated social systems, they must remember (without the help of language) who among them is dominant and who is not. Pre-linguistic humans may have faced similar challenges and solved them with mental recursion.

Certain ideas in the *Science* paper were familiar to anyone who followed Chomsky's work. He was, of course, the first linguist to attach importance to the fact that human brains can take a set of entities, such as words, and create an infinitely long pattern with them, such as a sentence. As we now recognize, this makes human language limitless, and most important, this recursive mechanism allows us to express complicated thoughts. We're not restricted to only one level of observation or knowledge; we can see (and say), "He knows," but also, "She knows that he knows." Each level of recursion is a step upward in complexity.

Moreover, Chomsky had previously suggested that the mechanism of recursion extended beyond language and was vital to human cognition more broadly. As the *Science* article pointed out, recursion is characteristic of the number system as well as the grammatical system. Just as "Mary thinks that" could be added to any sentence, "2x" could be added to any equation, no matter how long it already is.

In essence, the Hauser, Chomsky, and Fitch hypothesis said that although other animals may indeed have a rich understanding of the world, they have no way to convey it. It was only when humans connected their internal understandings with the means to express them that they gained their unique form of language. After the article was published, Hauser remarked, "When those things got married, the world was changed."

Steven Pinker and Ray Jackendoff published a response to Hauser, Chomsky, and Fitch, and a vehement back-and-forth ensued. (In all, four papers, including the original *Science* article, were published.) Pinker and Jackendoff charged Chomsky with having abandoned the last twenty-five years of his research and co-opting ideas from models he had once completely dismissed.

"I think the thing that startled a lot of people about that *Science* paper," said Jackendoff, "was that all of a sudden Chomsky seemed to be saying that language isn't so complex after all—that all this complexity is coming from the interaction of this very simple system with the interfaces, and so to many linguists it was like Chomsky was undermining the position on which we had all grown up and many of us still believe. Pinker's and my reply wasn't so much about the evolution of language as the character of language. We wanted to say, 'Look, there are all these complexities to language, and they don't reduce out to general capacities found in other animals.'"

Pinker and Jackendoff argued that Chomsky and his co-writers implied that Chomsky's linguistics was the only kind of linguistics there was, which in effect predetermined their definition of language. Throughout the paper, as throughout most of Chomsky's writing, language is described as having a "core"—a small set of very important features that lie at the heart of the phenomenon. But, Pinker and Jackendoff argued, there is no core to language. The appearance of one is just a mirage, an artifact of the way Chomskyans carve up language in the first place. Language is a complicated mass that can't be neatly reduced to a smaller concentrated essence or set of rules.

Pinker and Jackendoff also emphasized the idea of modifications taking place in organs and functions, in contrast to the Hauser, Chomsky, and Fitch hypothesis that traits could be assigned to one bin or another—broad and shared with many animals, or only human.

In an interview, Pinker later said:

> I don't think a theory of language evolution based on a theory of language that is idiosyncratic to one person's vision is productive. I don't think divorcing language from communication is a step forward, and I don't think writing off everything but syntax, indeed everything but recursion, and giving it to the animals, is a step forward. I think Chomsky so badly wanted to save something as unique to humans, namely the core of syntax, that he was willing to sacrifice everything else, in particular, the parts of language he is less interested in, like speech and words. It reminds me of the lizard that lets its tail break off when a predator is about to attack.

Philip Lieberman took the opposite view of the paper. "It's the same old Chomsky claim—a unique neural system or device specific to language exists in humans and humans alone, allowing infinite 'recursion.' It is a sea of words covering up Chomsky's unchanged view concerning the essence of language—it is a capacity shared by no other animal and distinct from any other aspect of human behavior."

For scholars like Lieberman, the authors' proposal to use comparative data to explore the question of language evolution was disingenuous. As he explained, "The comparative method has been used for many years to explore the evolution of language—my first published paper comparing monkeys to humans was published in 1968." Thus, rather than illuminate a way forward, the paper—for some of its critics—obscured the intellectual history of many of the studies it mentioned. Lieberman said, "The aspects of language that Hauser, Chomsky, and Fitch believe can be revealed through comparative behavioral and neurophysiologic studies are the ones that Chomsky and his disciples have always considered trivial and irrelevant."[2]

Similarly, William D. Hopkins, whose work with chimpanzees revealed Brodmann's area 44, observed that even though Chomsky was finally incorporating animal data, he was using it to designate commonalities between humans and other animals as somehow not important to language. "I'm not sure what that is," he said, "but it's not the comparative method."

As for recursion, Lieberman argued that it was adequately accounted for in the brain's control of the motor system. Pinker and Jackendoff pointed out that recursion occurs not only in language but also in vision, thus providing little motivation to restrict it to a narrow faculty of language. Irene Pepperberg noted that as far as comprehension was concerned, recursion isn't necessarily unique to humans. Still others raised the possibility that even humans don't do recursion either very much or very effectively.

Controversy over the paper has continued, and typical of the intense debates that Chomsky ignites, there is sometimes more emotion than accuracy about what is at stake. In one presentation at the Evolution of Language conference in Leipzig in 2004, the speaker, generative linguist Frederick Newmeyer, mentioned the article in an aside, remarking that he was bewildered by it. In response, an upwelling of muttering quickly turned into a shouting match. One researcher stood and shouted: "Chomsky says 'a miracle occurred.' Read it! He says 'a miracle occurred.' " Fitch was also in that audience. When he was able to get a word in edgewise, he said, "I'm a coauthor on that paper, and that word did not appear in it."

Today, the questions that remain most controversial in language evolution are the following:

- Was there *one* crucial gateway to language through which only humans have passed?

- Is there anything in the way language is processed by the brain that is unique to language, rather than a more general form of cognition?
- At what points in the trajectory of language evolution has natural selection come into play? Can any elements of the language suite be clearly identified as spandrels?

The first of the remaining questions reveals an odd, almost vestigial, way of thinking about the subject. We are aware by now that approximately twenty years ago language as a whole was seen as a single gateway through which humanity and no other extant animal has passed. In the face of the many arguments and experiments presented in this book, that idea has fallen apart. Language is not a single thing, and getting from no language to modern human language takes many steps. We are the only species alive today to have taken all of these steps—nevertheless, many other living animals have taken a considerable number of them (though not necessarily along the same path). Thus researchers like Irene Pepperberg talk more in terms of a rough continuum between modern animals and modern humans, describing the linguistic differences between them and us as more quantitative than qualitative. Such a continuum doesn't necessarily reveal genetic relatedness or trace evolutionary history, but rather is based on the existence of similarities and differences of features important to language.

Still, the idea of a single, categorical shift in the language evolution trajectory haunts the new field. In its latest incarnation, the debate is about whether we acquired recursion in a single move, and in doing so, language became what it is today and we became human, unique among all other living animals.[3] As discussed earlier, this notion has been objected to on several grounds, and many issues remain about how human-specific or language-specific recursion is, and indeed how often humans actually use it.

It's highly unlikely that a discrete feature could comprise the one, big difference between our language ability and that of modern-day chimpanzees, because their status as our closest living cousins is an entirely arbitrary one. We once had many closer relatives, and they have presumably gone extinct for a variety of reasons. Had the chimpanzee, bonobo, and gorilla gone extinct in the last century, our closest comparison would be with the orangutan, which would move the gap to an arbitrarily greater distance. Certainly no scientist has ever suggested that there is a single biological or logical reason for our current degree of uniqueness (or loneliness). Nor is there is anything significant about the human-chimpanzee split that led us to where we are now. Indeed, since our lineage split away from the chimpanzee line, it's overwhelmingly likely that our australopithecine and then hominid ancestors took yet more steps, moving through a number of forms of linguistic communication before arriving at the most recent stage of language—ours.

Dan Sperber, a social and cognitive scientist at the French Centre National de la Recherche Scientifique in Paris, makes an interesting case for a component of language that probably predates fully modern language but must have evolved after our ancestors split from chimpanzees. Sperber is well known for the theory he and the linguist Deirdre Wilson presented in a seminal 1986 book, *Relevance: Communication and Cognition*. Briefly, relevance theory holds that inference is as fundamental to linguistic communication as the ability to decode the words in a given utterance. For example, the sentence "It is too slow" may convey a variety of completely different meanings, given different contexts. Sperber lists a few of the possibilities: "The mouse is too slow in solving the maze; The chemical reaction is too slow compared to what we expected; The decrease in unemployment is too slow to avoid social unrest; Jacques' car is too slow (and so I'd suggest we take Pierre's)."[4]

Human communication in this view is about one person indicating his or her meaning to another. This can be done in a number of ways—via gesture, pantomime, or a linguistic code. The fundamental principle is that the person doing the listening (or watching, etc.) infers the speaker's meaning from the signal *and* the context in which it is conveyed. (The relevance of any message results from a shared set of assumptions. It is crucial, for example, that the listener knows that the speaker wants to convey a meaning, also that the speaker knows that the listener knows this; in addition, there is a shared assumption that what is communicated, regardless of the form it takes, makes sense within the context of the communication.)

Sperber and Wilson's theory effectively crystallized the intuition that the context of language really matters, and since then, depending on their focus, researchers have placed differing emphasis on the relative significance of the pragmatic aspect of an utterance versus its linguistic structure. Regardless of these differences, Sperber makes the point that all the linguistic sophistication in the world won't make language useful if its users are unable to infer the intentions behind an utterance and appropriately judge the relevance of its context. Likewise, Sperber points out that, compared to humans, chimpanzees have only a rudimentary ability to make inference about the beliefs and intentions of another chimpanzee. The ability to make sophisticated inferences about the relevance of a signal must therefore have preceded the final elaboration of structure in modern language, and it probably came after the split of our lineage with that of chimpanzees. [5]

It's not yet clear what type of investigation, experimental or otherwise, may further illuminate the relationship between pragmatics and linguistics in the evolution of language. Nevertheless, Sperber's broad point is that both must be explored. In addition to the ideas put forth by Terrence Deacon and Michael Arbib in chapter 14, he has offered an excellent candidate for a specifically human precursor to modern language.

If recursion did not transform our six-million-year-old grandparents into modern humans, perhaps it changed ancient humans into us by converting an archaic, simpler language into the version we have today. Is it possible that this is what happened two hundred thousand years ago? Yes. But if so, this shift is only one of many important turning points in the course of language evolution.

Ultimately, the notion that a single attribute will explain why humans are the only living species to have language is as unhelpful in its latest version as in its oldest. There are hundreds of gateways to linguistic communication, and the evolutionary process provides no motivation to hail *one,* distinct from all the others, as more integral to language. The problem "Is there *one* crucial gateway to language through which only modern humans have passed?" may still be much discussed, but in all of its forms, it is truly a nonquestion.

If good science doesn't focus on *one* stage in linguistic evolution at the

expense of all others, it will inevitably highlight only a selection. This is because some steps will be more experimentally tractable, while others will
be easier to observe. Some steps will be notable because they preceded or
followed a dramatic cultural shift.[6] Some may be considered research-
worthy because of their extreme remoteness in time, because they result
from a human-specific genetic mutation, or because they drove the selection of a relevant mutation. If a stage in language evolution were ever
linked to one of the few genes unique to *Homo sapiens,* it would draw enormous interest. Naturally, some steps will just seem more interesting because of what we think they imply about us.

In the current debate, even though different researchers talk in terms
of continuity and discontinuity or qualitative versus quantitative differences, there is nevertheless a greater and more important convergence on
the same data and some basic concepts. To a large extent, the conflicts
noted here are characterized by different emphases and focus rather than
by completely opposed ideas. The argument between Pinker and Chomsky
and their coauthors about FOXP2 illustrates this rather well. Pinker and
Jackendoff argued that the importance of FOXP2 is that its sequence is
uniquely human. Chomsky, Hauser, and Fitch discuss FOXP2 in very different terms—the gene that subserves language is shared by many different species and is therefore likely evidence of the broad foundations of
language.

Both are right. The shared nature of the gene implies an ancient history
and widely dispersed potential for development along the language path.
Nevertheless, a uniquely human mutation of FOXP2 has been positively
selected in our species within the last 200,000 years. The FOXP2 mutation
is a significant twitch on the genetic dial that accompanies the emergence
of human language. Beyond this, the individual researcher may decide
what matters most to him—the dial or the twitch.

As for whether there is anything unique to language in the human
brain, the question becomes complicated by the need to consider the
development of the individual, the development of the species, the way
that language itself changes through time, and the way that all of these factors interact. What's certain is that the question no longer makes sense in
the terms in which it used to be posed, that is: Is there a specific gene that

programs a specific chunk of the brain to be a language processor? Nevertheless, it does appear that language doesn't just fall out of the adult human brain without some specifically linguistic processes occurring, as shown by the neural route taken by regular past-tense verbs in contrast to irregular ones.

What about the evolutionary processes of adaptation, where a trait evolves for a particular purpose, and exaptation, where a trait that is used for one function becomes co-opted to serve another purpose in later generations? What role have these played in language evolution? For all the furious words expended on the subject, everyone agrees that both processes have had a role. And everyone has acknowledged that communication has to have something to do with language evolution. Regardless, the rapid spread of the human mutation of the FOXP2 gene is definitive evidence that there has been positive selection for a form of gene that had major consequences for language.

It's not just the genetics that make the spandrel suggestion unlikely. Humans accumulate a great deal of knowledge in their lifetimes. They are also an extremely social species. Could it just be a coincidence that we are able to communicate all that knowledge to other humans? "We're social creatures," said Pinker. "We don't just cooperate with our kin, we negotiate agreements with people that we're not related to, and societies are formed by implicit social contracts and exchange and understanding. If language was really just a by-product, one wonders why there would be such an amazingly good fit into the rest of what makes us zoologically unique."

Many exciting angles remain to be further explored—for example, what's essential to language development and what is helpful but ultimately incidental? Language clearly bootstraps itself from gesture, but does a species have to have gesture to develop some form of language? How many individuals do you need in a species, as well as in a community, for language to arise in the first place and for it to be passed down through the generations and keep evolving itself? If you have to be human to have human language, could another species in different conditions ever evolve a form of language that used enough of the same basic building blocks that we could translate between our language and theirs?

The jury is out on these questions, though there is every reason to believe that the more data that are generated, the closer we will be to an answer. We can expect resolution on how powerful an evolutionary force communication has been and what elements of language it has shaped. In addition, we can hope to know more about how fine-grained the back-and-forth of modification and selection has been. Were some spandrels adopted as a piece into language? Or did some small increase in the power to compute a grammatical relationship arise as a spandrel and then become further elaborated over a long history of adaptation?

If there were a moral to the story of evolution, it would be that meaning is something that happens after the fact. There is no rhyme or reason to the mutations that occur over the evolution of a species. Within the constraints of what has so far developed, genetic mutations are random; it is what the creature does with them that makes them meaningful. Evolution is the opposite of destiny, and because we are creatures of both biological and cultural evolution, where we are going is really obvious only in hindsight.

Certainly, it's impossible to predict the future of Chomskyan influence. Chomsky is most famous in cognitive science for being the first to point out that language is both extremely complicated and innate. Now the main complications are how language is defined, what the goals of scientific endeavor are, and the strange and enormous sociological phenomenon that Chomsky has engendered.

Within the field of language evolution, Chomsky is associated with the caveat that language may have as much to do with inner speech as it does with communication between two individuals. But the value of this caution is questionable. We all have the sense that words exist inside our heads and that this sensation accompanies thought. But what forms do the words in one's mind take? How complete or incomplete are mental sentences? How could so subjective an experience even begin to be measured? No researchers have been inclined, or able, to take the basic idea any further than the form in which Chomsky first suggested it.

Indeed, though Chomsky has thrown out this possibility on a number of occasions—and although he is interpreted by many as saying that this is why language evolved—on other occasions he has qualified it further. In

2000 he wrote, "One can devise equally meritorious (that is, equally pointless) tales of advantage conferred by a small series of mutations that facilitated planning and clarification of thought . . . not that I am proposing this or any other story."[7]

Chomsky's focus on extraorganic principles and the idea that we just don't know what happens when you pack that many neurons into a space that size is an important part of the debate. Recently a number of mathematicians at the Cold Spring Harbor Laboratory in New York announced that they had worked out the mathematical basis for why the brain is divided into white matter and gray matter.[8] This is exactly the kind of idea that Chomsky has been promoting since the 1970s.

Still, the way Chomsky carved up the linguistic universe is unacceptable to many researchers in language evolution. Even Ray Jackendoff proposes dismantling the long-standing Chomskyan ideas that the complexity of language arises out of the complexity of syntax and that syntax is central to language. Many researchers over the years took extremely seriously the idea that syntax was autonomous and somehow preexisted everything else in language. Gallons of ink have been spilled in the attempt to build models of a language processor that contains a separate syntactic processor, which analyzes the abstract structure of spoken language even before the sound. But, says Jackendoff, it's time for this to be discarded.

Typically, Chomsky has been ambiguous, enlightening, and dismissive of the new ideas about the emergence of language. For example, Terrence Deacon's book *The Symbolic Species* was received with admiration by many in the field. Chomsky, on the other hand, wrote, "I have no idea what this means." Deacon's account of linguistics, according to Chomsky, is "unrecognizable." He concluded: "I do not recommend this course either; in fact could not, because I do not understand it."[9]

One striking effect of the paper that Chomsky co-wrote with Hauser and Fitch was that it seemed to make other researchers in the field even more sensitive to, and critical of, Chomsky's vast influence. Derek Bickerton, who years before had written that nothing really happened in linguistics before Chomsky, wrote about the Stony Brook conference on his blog:

On October 14, 2005, Chomsky disembarked on Long Island for one of the few conferences he has attended in the last several decades: the Morris Symposium on the Evolution of Language at S.U.N.Y., Stony Brook. He arrived too late for any of the presentations given by other scholars on that date, gave his public lecture, gave his conference presentation at the commencement of the next morning's session, and, despite the fact that all of the morning's speakers and commentators were expected to show up for a general discussion at the end of that session, left immediately for the ferry back without having attended a single talk by another speaker. For me, and for numerous others who attended the symposium, this showed a lack of respect for everyone involved. It spelled out in unmistakable terms his indifference to anything anyone else might say or think and his unshakable certainty that, since he was manifestly right, it would be a waste of time to interact with any of the hoi polloi in the muddy trenches of language evolution.[10]

Does the fact that Chomsky is now contributing to the discussion on language evolution mean that he is conceding it is crucial to linguistics? Pinker said no. "He gives with one hand and takes with the other. Chomsky says, 'All hypotheses are worthless, so here's mine, which is as worthless as anyone else's.'" This latest gyration in a long career of twists and turns, Pinker said, marks the beginning of Chomsky's decadent phase.

What does it mean that one man had such a long-standing and wide-reaching impact? "I don't think it is good," said Pinker.

Because Chomsky has such an outsize influence in the field of linguistics, when he has an intuition as to what a theory ought to look like, an army of people go out and reanalyze everything to conform to that intuition. To have a whole field turn on its heels every time one person wakes up with a revelation can't be healthy. It leads to a lack of cumulativeness, and an unhealthy fractiousness. It's an Orwellian situation where today Oceania is the ally and Eurasia is the enemy, and tomorrow it's the other way around. Time and effort and emotional effort get wasted.

Ray Jackendoff likened Chomsky's persona and influence to that of Freud in psychoanalysis. "Freud especially is an interesting model," Jackendoff said. "Even though the specifics of the way Freud thought about things have been shown to be incorrect, nowadays we still take for granted all the basic ideas of Freud's approach to the mind, about people's motives and what drives them. Everyone who goes to a therapist now owes it to Freud. The same is true of Chomsky. The idea that you can look at language as a computational system invested in the mind and that there's an acquisition problem that requires some question about what the child is bringing to the learning process, and that there are formal tools for discovering language in great detail—that's now taken totally for granted in the field and that came from him."

The study of language evolution is in some ways the opposite of the formal linguistics that Chomsky created. It doesn't start with language as a formal abstraction, but grounds it first in the human body, and in history. The questions that Chomsky considers critical, such as "Is language useless but perfect or useful and imperfect?" are not much discussed outside considerations of his own work. As for the notion that linguistics poses a crisis for biology, most evolutionary biologists and other researchers in the field seem confident that they can be brought into consilience.[11]

The power that Chomsky has wielded and still does is impressive. Many researchers regard the ideas in the *Science* language evolution paper as just the natural maturation and progression of a brilliant mind. This one man and his unique ideas have influenced literally thousands of academics. In the early days of language evolution, his name was used as an obsessive touchstone in many articles. But people now seem to be freeing themselves from that influence.

Few are up to the task of disentangling the ideas *attributed* to Chomsky from the ideas that *really are* his. Without a doubt, people hold him responsible for things he didn't say. And he is often accused of denying things he did say.

As for generative linguistics, in the gentle phrasing of Jim Hurford, it is taking the burden off universal grammar. Indeed, all the evidence about genes, gesture, speech, physiology, and brain damage point away from UG.

Today, many researchers who argue that the innateness of language is neither language-specific nor grammatical in nature still use the term "UG." Some researchers even go to the trouble of pointing out that what they mean by "UG" is neither universal nor a grammar, a caveat that surely qualifies the term as either misleading or irrelevant.

Only time will tell if the magnitude of Chomsky's influence will persist. Currently the divide between his many critics and supporters remains religious in its zeal, with many researchers believing that Chomsky is an academic villain who led linguistics completely astray. In some lights, however, their problem is a definitional one. Chomsky's interest extends only to what he considers the syntactic core of language. This necessarily excludes all this other study. Why should this matter so much? Having interests, and therefore areas of indifference, is a freedom allowed most everyone else in academia, but Chomsky's lack of interest in a topic often leads to umbrage. Others still see him as the source of everything we now know. Charles Yang, a professor of linguistics at the University of Pennsylvania and author of *The Infinite Gift,* wrote in the *London Review of Books* that Pinker and most other researchers are merely turning over the rocks at the base of the Chomskyan landslide.[12]

As for language evolution, these facts are undeniable: Chomsky dismissed it for a long time, his dismissal was treated as an irrefutable argument, and now language evolution has taken on a life of its own. Probably the truth is that the boom in language evolution has occurred both because of and in spite of him. Chomsky brought the attention of the world to the complexity of language and its innateness. Whether his version of complexity and innateness will endure is another matter.

The overriding outcome of the language evolution debate kicked off by Chomsky's 2002 paper was that it became abundantly clear to everyone in the field that, as Jackendoff put it, one's theory of language evolution depended on one's theory of language. And even though Chomsky's contributions set the agenda for linguistics and cognitive science for the latter part of the twentieth century, many researchers rejected the way that that paper attempted to rein in all the evidence and set the crucial questions for language evolution in the coming century. There's no doubt that Hauser's and Fitch's experimental work will be central to the ongoing language evolution dialogue, but the specifics of the Chomskyan framework may not last as long.

In some ways, it bodes well for the study of language evolution that it can't yet be compressed into a neat framework. It has always been a quirky field, and it retains much of its oddness. For example, the energetic back-and-forth between Pinker, Jackendoff, Chomsky, Hauser, and Fitch belies the fact that all five subscribe to a basic model in which language is somehow generated from the human brain.

Lieberman, on the other hand, is antigenerativist, and yet both he and Pinker agree on a first and fundamental principle—that you have to start with evolution in order to really get at the true nature of language. Jackendoff, who has been a Chomskyan linguist from the very start of the Chomskyan era, now proposes that formal grammars should be constructed so that they are consistent with an exploration of language evolution.

Within language evolution, computational modeling has been an enormous hit. In fact, Simon Kirby's success with modeling has led him back to an interesting place. Now he's trying to run generations of language learning through the minds of real people. He recently conducted a pilot study where he put individuals in a room and presented them with a small-world, talking-heads-style experiment. The subjects looked at a screen that contained a number of objects that were distinguished along a few dimensions, like color, shape, and movement. Across the bottom of the screen ran a series of invented words, an "alien" language that described what was pictured on the screen. The subjects were asked to try to learn the alien language. They were then tested on a series of pictures, which included some they hadn't seen before (hardly any of the participants noticed this fact). Inevitably the subjects did not feed back only the language elements that they had been given. There were mistakes, modifications, and elaborations.

The study was intergenerational, because Kirby ran the subjects one after the other, and each time the alien language was, in fact, the answers the previous subject gave to the test pictures. Except for the initial random language given to the first subject, there was no alien language, only the contributions of each individual, which were culturally transmitted from generation to generation. Each subject in the experiment believed that he was simply giving back what he had learned, but instead the language was evolving. "It's the same as the modeling," Kirby explained, "in that it gets easier to speak the language with each generation." He had originally thought that speakers might generate different elements to mark each of the features and then combine them in a precise kind of way. But that's not how they did it. "People take whatever elements of structure they are given," said Kirby, "and they go with it."

Kirby's first foray into modeling language evolution with human agents bears out what his digital models have predicted. "Structure organically emerges in the alien language, and it does it in a cumulative way. No single individual has created structured language, but it emerges after several generations from the accretion of lots of individuals' contributions." Darwin alluded to the emergent properties of language when he wrote in *The Descent of Man* that language is a cultural invention, though *not* a conscious one. As he and others have put it, the appearance of design does not necessitate the work of a designer. Kirby said, "It is real cultural evolution, steered by the biology of our experimental participants, but with an evolutionary dynamic and adaptive logic of its own. Features of the evolving languages in our experiments are there for their own selfish reasons (they are better at surviving to the next generation), not because of our desire to invent them."

Luc Steels's work heads in ever more creative directions. Steels, Vittorio Loreto (a physicist at the Università di Roma), and other colleagues are investigating ways to integrate what is known about the dynamics of semiotic systems with technology. The researchers are intrigued by the way that Web sites such as del.icio.us and flickr.com enable users to tag online resources, sharing commentary and other data with users. "Tagging sites glue online social communities by pushing thousands of people to take part in a collective effort to attach metadata," said Loreto. With these sites, the popularity of a tag will typically begin to spread slowly; however, there is a phenomenon where one tag may suddenly become significantly more popular than all the rest. Steels and Loreto's new experiments with autonomous agents engaged in language games (such as Steels's "talking heads") are showing that in the same way that widespread agreement about a tag may suddenly emerge in a social networking site, there can also be dramatic transitions in a network of digital agents, where a shared set of conventions suddenly replaces a phase of chaotic disagreement. The dynamics of meaning can help explain a similar phenomenon in human communication—how large populations of speakers suddenly converge on the use of a new word or grammatical construct.

This has obvious implication for stages of language evolution, where a new level of complexity replaces a previous level without any conscious agreement by protospeakers. Loreto and his colleagues suggest some interesting ways to exploit semiotic dynamics. For example, scientists could deploy groups of robots with such capabilities in situations where contact with humans is unreliable or impossible. Such robots might explore distant planets or deep seas, creating a way to communicate about, and respond to, events that were completely unforeseeable by their human programmers.

The involvement of a physicist like Loreto in a project connected to language evolution is a striking sign of just how many tentacles this problem has. Another language evolution researcher with a surprising background is Ramon Ferrer i Cancho. He is a former computer scientist who now works in the Department de Física Fonamental at Universidad de Barcelona. Ferrer i Cancho uses Zipf's law to model language, exploring the trade-offs between speakers and hearers during communication.

Speakers must make an effort in order to be understood. For a speaker to be as clear as possible and avoid ambiguous meanings, greater effort is required. Listeners, on the other hand, must make an effort to interpret the correct meaning of an utterance, and they must work harder to decipher the intent of a speaker who has devoted less effort to clarity. Accordingly, Ferrer i Cancho's models explore what happens when there are small shifts in the balance between the effort of the speaker and the hearer. In fact, a tiny change in the balance between the two can dramatically alter the properties of a communication system. Says Ferrer i Cancho, it's possible that similarly small changes may underlie a dramatic shift from a communication system with a simple vocabulary made up of a few precise words to a larger vocabulary with varying levels of semantic precision.

The history of animal language research has been a turbulent one, but that may also be changing. Of language evolution conferences, Heidi Lyn said, "If I go to talks by some of the more established people, it tends to be either that they don't mention the ape language research at all or they dismiss it. And there are people who consistently stand up and get things wrong. For example, an older linguist at the Harvard language evolution conference in 2002 who was asked about Kanzi dismissed him. 'Kanzi's an aberration,' he said. 'He is the only example that we've ever seen of this.' " At the same conference, Herb Terrace stood up and asked Lyn if Kanzi was trained with food rewards. Lyn explained that they didn't do this, yet Terrace persisted with that line of questioning. "It's different with scholars my age or younger," Lyn observed. The next generation gives a lot more credence to ape language research, and to work like Sue Savage-Rumbaugh's. "They are willing to look at the data," said Lyn. "It's not just a matter of age. It's the difference between people who lived through the Terrace criticism and the people who didn't."

For more than two decades Savage-Rumbaugh herself has been working closely with scholars from a language research program in Atlanta to apply the picture keyboards and other techniques she has used for communicating with the bonobos to communication with mentally retarded individuals whose levels of language skills have reached only those of small children. They have had great success with some individuals, equipping them with an ability to connect with other human beings that they wouldn't have otherwise had.[13]

Other applications of language evolution research are completely futuristic but, at the same time, surprisingly practical. Philip Lieberman's experiments on Everest not only illuminate the path that language evolution took but are serving as a model for NASA to monitor the well-being of astronauts on their way to Mars. The brain damage that Everest climbers suffer when they experience oxygen deprivation is similar to the kind of damage that a Mars-bound astronaut would incur from exposure to cosmic rays. If scientists back on Earth are able to detect subtle or profound neural damage in astronauts simply by listening to how they pronounce certain vowels and consonants, they'll be able to react, and, it is hoped, treat them accordingly. This same project is also promising to improve the early diagnosis of Parkinson's disease, not to mention help the mountain climbers of the world.

The way that evolutionary research is redefining language has social consequences as well. Lieberman argues that if language were a true instinct, if it simply flowed from every single one of us regardless of the environment into which we were born, then our governments would have very little responsibility to promote its expression. Because language is a skill, and one that is closely connected to thinking, he says, it is improved by practice and training and environments that are conducive to learning. This creates a civic responsibility to help all students hone their language skills.

At the Evolution of Language conference in Rome in 2006, Tecumseh Fitch listed the many ways in which the field had made progress since the 1866 ban on the subject. He started by noting that for the first time at the language evolution meeting, no one had mentioned the ban.

16. The future of language and evolution

Five years after Pinker and Bloom wrote about the evolution of the eye and its lessons for language evolution, Dan-Eric Nilsson and Susanne Pelger of Lund University in Sweden published a paper called "A Pessimistic Estimate of the Time Required for an Eye to Evolve." Nilsson and Pelger digitally modeled the trajectory of the eye, beginning with a flat light-sensitive patch of cells—the kind of simple eye that we know some creatures have—and inflated it over time into a fully functioning mammalian eye.[1]

The scientists worked out a sequence of very small changes that had to occur if the light-detecting cells were to evolve into the separate specialized parts that interact with one another in an eye. For their model to be realistic, each small evolutionary step had to confer some survival advantage and therefore improvement in vision. Even though the changes were extremely tiny (no more than 1 percent change at any one time), each slightly modified eye was able to detect more and more spatial information. As the title of the paper suggests, Nilsson and Pelger erred on the side of pessimism, always assuming that it would take more generations for the eye to evolve rather than fewer. Given this, they calculated that it would take about 1,829 separate evolutionary steps for the flat-patch eye to evolve into a stereo-vision globe. That amounts to less than 364,000 years, not long at all from an evolutionary perspective.

We know from the fossil record that animals with modern eyes lived as early as the Cambrian period, 550 million years ago, which means there has been time for eyes to evolve more than fifteen hundred times since then. As perfect and wondrously complicated as our eyes seem to us, they are not irreducibly perfect from an evolutionary perspective.

To extend Pinker and Bloom's analogy to language: this means that abilities and organs that seem wildly complicated from our perspective may be able to come together relatively rapidly as functioning, complex wholes. In addition to this biological potential, we know from the work of people like Deacon, Kirby, and Christiansen that language itself may also evolve and that linguistic evolution occurs even more rapidly than biological evolution. Language may have appeared very recently in the human lineage, but that doesn't mean it was the product of a single, crucial event. No one mutation of genes or social order caused language to erupt from the mouths of our ancestors.

Even if researchers can't pinpoint every evolutionary event that led to the language we have today, and even though we don't know exactly what all the bends in the historical road looked like, the principles for further illuminating the path of language evolution are now self-evident. Fundamentally, the appearance of design in biology and in language can be taken as a sign of evolution, not of a designer. Additionally, where complex design does exist, it makes sense not to treat the whole as a monolith that simply developed from nothing to something in one or two quick steps. Finally, the most likely scenario is that both evolutionary novelty and derivation played a significant role in the evolution of a phenomenon as complex as language.

What does it mean that we are getting closer to the answer of how language evolved? The implications are as diverse and varied as the story of evolution itself. First, from a research perspective, it means that good data lead to better data, and there is still a great deal of data to be gathered before the big picture can be filled out. "People have been arguing about Neanderthal speech for the last thirty-five years and whether chimp sign language is really language," said Tecumseh Fitch, "yet nobody even thought to ask what chimps do when they vocalize. We still don't know—nobody's put a chimp in an X-ray setup and watched it vocalize. It's amazing how much data is out there that hasn't been collected, like taping birdsong and whale song and doing linguistic analysis of that. We could apply this huge theoretical apparatus that phonologists have developed to birdsong. It's not even that hard, and it's an obvious thing to do." Fitch added, "What amazes me coming into this field is how many things you can answer that no one even thought to look at."

One of the biggest questions yet to be answered is posed by Ray Jackendoff: How do neurons do it? Magnetic resonance imaging and other ways of seeing the brain in action have taught us a lot about how our brains function. Overall, imaging has shown that for many higher-level activities, like language, neural activity is distributed across the brain. There are no specific areas that light up for language and language alone. Still, there's no doubt that scientists fifty years from now will find the wonders of our neuroscience to be fairly crude. Although we can now map the brain as it works, we still have no actual idea *how* it works. How do the neurons do what they do? How do they process, store, and produce language? There is no predetermined meaning inside our heads. Neurons don't contain symbols, but mainly pass on (or don't pass on) activation signals to one another. So how can the patterned flare of electrical charge across our brains mean that we recognize the word "cat," even when it is spoken by one hundred different speakers with their one hundred unique voices? How can we tell the difference between a *p* and a *b* when there is no tiny prototype of these sounds deposited in our neurons?

"We know we can't think of the brain as a digital computer anymore," said Jackendoff. "It's sort of a parallel, semi-analog computer. But how does it do these digital things?" Discovering how neurons work should allow us to determine once and for all which of these frameworks for analysis—from the prototypical *p* to the syntax of English—are real and which are mirages.

It's clear by now that the problem of language evolution is completely intractable when you approach it from the perspective of a single discipline. For all the salient questions to be answered, the multidisciplinary nature of the field will have to become even more so. So far, it has taken years for individuals in different departments to start talking, to develop research questions that make sense for more than one narrow line of inquiry, and to start to understand one another's points of view. The field of language evolution needs students who can synthesize information from neuroscience, psychology, computer modeling, genetics, and linguistics. The more this happens, the richer and wider the field will become, instead of devolving around one or two theoretical issues.

Technology and wide-ranging discussion are not the only factors that will aid the next big leaps in understanding. Much of the impetus will come from the fact that a generation of scientists has broken free from the iron grip of some old ideas, while other notions that were once regarded as radical, or at least unpopular, have spread into the mainstream in all branches of science. The notion that animals do not think—or that, if they do, it is completely and qualitatively different from human thinking—is finally dying, if not completely dead. This idea shaped research in many different fields for decades, both in a direct way and by scaring people off the topic for fear of looking foolish.

The flip side of the animals-are-dumb belief is the idea that human thinking is boundless and that our language is infinitely expressive. Yet evolutionary theory, which tells us, first, that we are a particular type of creature, not an über-creature; second, that our brains are particular types of thinking machines, not all-purpose thinking machines; and, third, that although the structure of our language means we can be extremely creative, we are only as likely to express infinite meaning as we are to talk for eternity.

No matter what their particular take is on complexity or innateness, most theories of language and evolution have one thing in common: they focus on what's happened in the past up to the present. It's an obvious frame of reference, but sometimes that focus gives the impression that the present is an eternal moment that will stretch forward into the future, with us—and language—remaining unchanged forever. Some scientists have even argued explicitly that we have stopped evolving.

Certainly humanity is a powerful force of selection, both on other species and on ourselves. We have been manipulating the genomes of plants for thousands of years in agriculture, and we've been doing the same thing with livestock, as well as with dogs, cats, and other domesticated animals. The sheer weight of the human biomass and all of its accessories—its buildings, fields, roads, dams, and cell phone towers—affects the survival of other species by pushing them into smaller and smaller niches. We deselect the genomes of some animals, like the mammoth, by hunting them to extinction, and we pollute, poison, and inadvertently engineer the genomes of others—like fish whose DNA is corrupted by human estrogen in waterways. We introduce alien species into new environments, where they decimate local populations or rapidly evolve themselves in order to survive. Our use of pesticides and drugs induces the ultra-rapid evolution of resistant strains of bacteria and viruses. And of course we change the natural history of the human genome with the mass production of food, medicine, and health care. Diseases and traumas that would otherwise kill us before we had a chance to reproduce can today be completely averted. Similarly, men and women who would otherwise not be able to conceive can now bear children with the assistance of reproductive technologies. In fact, a generation of children whose parents were among the first to undergo in vitro fertilization are now a far-flung group of young adults bearing their own children and spawning a generation that in another time could never have existed.

Today humanity is tinkering inside the evolutionary machine itself, altering DNA directly. Normally, in the shuffle and flow of evolutionary change, no single genome occurs more than once—except, of course, when twins or other multiples are born. But in 2006 we cloned cats and dogs for the first time, and these animals were just the latest in a growing list. No one can reasonably expect that a cloned human is far off. We're also tinkering with the ways genes express themselves in individuals. The intent behind this science is not just to head off illness but, for some researchers, to bioengineer designer human beings.

While all living things affect the evolution of other living things simply by virtue of trying to stay alive, humans interact with the biological evolution of other species in a much more complex and powerful fashion because of one ability: language. Nothing occurs on the human scale without language. No language means no agriculture, no animal farming, no science.

Still, as fascinating and unprecedented as this moment in the history of life on earth is, it is only a single point in time. We tend to assume that our current evolutionary stage is the inevitable endpoint of some natural drive to complexity and intelligence, but *now* is merely an arbitrary instant. The future stretches out before us, and, as the saying has it, it's going to go for a lot longer than the past. As far as our species is concerned, this "modern" era may well be the dawn of time. Certainly, the fossil record reveals that anything can and does happen. Ice ages, meteors, killer viruses, and tsunamis occur and recur, and these are only the most dramatic and obvious events that can alter the course of a species—either by selecting some genomes over others or by extinguishing them entirely. The only real measure of success on this planet remains what it has always been: not language, but life. Our species survives. And every other type of animal that doesn't possess human language but still exists, by definition, also survives. The notion that we may have halted evolution or stopped evolving ourselves is just another version of the seductive but empty idea that we have control over our destiny, either as individuals or as a species.

In 2005 scientists published the results of a number of experiments that indicated that humans are still evolving. In one case, a team of geneticists led by Bruce Lahn at the University of Chicago offered proof that the human brain has been continuously evolving since *Homo sapiens* first appeared. The scientists looked at two genes known as microcephalin and ASPM, both of which are known to contribute to brain growth.[2] (They are also expressed in other tissue in the body.) The geneticists sequenced DNA from a collection of human cells that represents the variation in our species, and they found that one variation of each gene, called an allele, occurred with particularly high frequency. The fact that the alleles seemed to occur more than normal genetic drift would allow suggests that they have been actively selected over time. The scientists believe that the frequent allele of microcephalin appeared around thirty-seven thousand years ago and the frequent allele of ASPM appeared only fifty-eight hundred years ago. It's not known what effect these versions of these genes have, or why they were selected. They could have shaped cognition, as Lahn argues. Other scientists suggest the genes could have had some other effect on the brain that doesn't directly impact thought.

At the same time that Lahn's results were published, another team of scientists based at the University of California, San Diego, announced the discovery of a positively selected gene called SIGLEC11 that is expressed in brain cells called microglia. Although they can't yet explain the effects of the gene, it is interesting because it is one of the very few found only in humans and not in our ape cousins. This could make it a candidate for explaining some of the differences between us and them.

Another direct case study of natural selection at work in humans today is an experiment carried out by scientists in Sweden. The study showed that a chromosome with a particular arrangement known as an inversion is positively selected for in the people of Iceland. The inverted form is one of two possible arrangements of the chromosome, and it occurs rarely in other human groups (hardly ever in Africans and virtually never in East Asians). Nevertheless, it is carried by 20 percent of the population of Iceland, and the women who carry this particular form of chromosome have more children than those who do not.[3]

The two possible arrangements of the Iceland study chromosome are known as H1 and H2, and they are thought to have split from the original chromosome three million years ago. As findings like these accumulate, they reveal not only that evolution has not stopped but that we are necessarily creatures of time. We could never have existed in our current form three million years ago—and if the evidence for ASPM is correct, we didn't even exist in the same form only ten thousand years ago. From gene to chromosome to different kinds of gene expression, human beings are as changeable as all that. In a 2006 study, the geneticist Jonathan Pritchard and his colleagues at the University of Chicago announced that there were at least seven hundred regions of the human genome that had clearly undergone positive selection in the last five thousand to fifteen thousand years. Some of the genes affect taste, smell, digestion, and brain function. It is thought that some of these changes resulted from the pressures involved in moving from a hunting-gathering lifestyle to a more agriculture-based one.[4]

Not all change is good. As much as language enables us to control nature and keep our environments stable, it also makes possible the dramatic altering of our environment in unexpected and dangerous ways. The same language skills that promote technological innovations like water irrigation, road building, and air-conditioning also produce the ozone-destroying pollution and countless other ecological dangers of the modern age. Any of these phenomena could result in a sharp left turn for the human genome. And perhaps the same linguistic skills that give us science, and currently some control over DNA, will lead to our own extinction in less obvious ways. Language and material culture have greatly increased the mobility of the world's population, and some researchers believe that this will lead to an unhealthy and irreversible diminishing of variation in our genome. As more and more humans breed across the boundaries of genetic variation, we become a blander, more homogeneous bunch than our diverse parent groups. This could be a problem because variation is important to the evolutionary health of a species, for the more we are the same, the easier it is for one single thing to make us extinct. Indeed, some genetic variants of the human species are disappearing altogether as small indigenous groups die out.[5]

Freeman Dyson, a well-known writer and retired professor of physics at the Institute for Advanced Study in Princeton, New Jersey, argues that one day in the not too distant future, biotechnology will become widely available to all. Gardeners will use do-it-yourself kits to engineer the plants of their dreams, and hobbyists and animal lovers will directly tinker with the genome of their favorite animal. Dyson thinks that children will also have access to toy genetic kits in much the same way his generation played with Erector sets. "When teenagers become as fluent in the language of genomes as they are today in the language of blogs," he writes, "they will be designing and growing all kinds of works for fun and profit."[6]

Is it possible that even if we have not stopped evolving, language itself has?[7] Mainstream linguistics assumes that language has hit a steady state, and that even if words and phrases appear and disappear—indeed, even if there is a change in the way fundamental roles like actor and object are marked—language remains essentially the same. And yet the linguistic landscape appears to be a rapidly changing one. Today there are about six thousand languages in the world, and half of the world's population speaks only ten of them. English is the single most dominant of these ten. British colonialism initiated the spread of English across the globe; it has been spoken nearly everywhere and has become even more prevalent since World War II, with the global reach of American power. Currently about 400 million people have been born to speak English, and another 430 million have learned it as a second language. (It is the most popular language for students of a foreign tongue.) But even its commanding dominance doesn't mean English will always be the world's most spoken language, and experts even doubt that it will be the chief language of the near future.

It's not yet possible to say which of the large and complicated currents that move through the world's languages are indicative of evolutionary change or just change. Within languages, some linguists see signs that evolution is afoot, such as John McWhorter, who argued persuasively that all languages are not the same because they are not equally complex. Perhaps this is the kind of variation that future moments in evolution will act upon?

Linguists who take a functionalist approach to grammar argue that the complexity of a language is shaped by the needs of its speakers rather than an innate grammar module. It is these relatively universal forces, they say, that mean some languages are more or less complex than others. This implies that grammatical structures arise in a language only as required by its speakers. Joan Bybee describes how languages that are historically and geographically unrelated undergo syntactic change in very similar ways—for example, verbs meaning "want" or "go" may become future tenses (as in English, "wanna," "gonna"), and the numeral "one" can turn into an indefinite article (as in English "a/an," German "ein/eine," French "un/une," and Spanish "un/una").

A pure functionalist would find little that is linguistically innate in humans, while an extreme nativist position finds almost everything innate. Taken as a whole, the data presented in this book support neither end of the continuum. Instead, they are compelling evidence that human specialization for language exists, and that forces that have often been neglected, such as the needs of speakers to communicate and indeed the need of a language to survive, contribute to the dynamic character of human language and to evolutionary change.

Extinction is as fundamental to the big evolutionary picture as survival, and certainly the extinction of languages continues all over the world. As human groups perish or shift cultural and political allegiances, their languages die too. The world loses one of its six thousand languages every two weeks, and children have stopped learning half of the languages currently spoken in the world. It's been argued that languages are under greater threat than any endangered bird or mammal.[8] Whether or not it's moral to let language extinction occur, it is the case that languages are irreplaceable records of the development of human societies and alternate windows into the human mind. When a language dies, we lose the knowledge that was encoded in it. Though we assume that when knowledge is lost, it has been superseded by a superior version, a dead language, with all its unique ways of carving up the world, is as irreplaceable as the dodo and the *Tyrannosaurus rex*.

Unfortunately, even if we, and our languages, are still evolving, we still don't know where we are heading. Things will probably remain unchanged for quite some time, and then . . . they won't.

Kurt Vonnegut wrote about the end of the world in *Galápagos*. In the novel a global disaster kills off most of the human species, but one small group survives, washed up on the Galápagos Islands. As time passes, evolution works its magic on the survivors' descendants, and traits that are not conducive to survival are inevitably superseded by those that are. In Vonnegut's brave new world, big brains are no longer an advantage, but a sleek, powerful swimming body is, and *Homo sapiens* end up seal-like and simple.

Vonnegut exposes the assumption that if we do change biologically, we typically think we will end up smarter in the terms in which we consider ourselves smart today.[9] But to survive means only that we'll be smart in the context of the environment we find ourselves in. If we continue to exist, we will by definition be smarter than the versions of us that did not survive, but that intelligence won't necessarily be comparable to what we have today.

At least individually, we do know where we are going: you and everyone you know are going to die. For this awareness, you can thank language. Talk about spandrels.[10] The same linguistic structures that allow us to soar through time and space and model entire universes in our heads also enable us to foresee our own mortality. Language also permits us to imagine a self that isn't earthbound and a world beyond death. So far it hasn't offered a way to avoid it.

Scientists typically offer up the wondrous metaphysical architecture we build with language as a consolation for our mortality. We may not be here for long, but because we have language, we can understand the way that the cosmos spins and twists back on itself, we can see the scintillating and sticky interplay of all the particles of existence, and we can work out the way that small evolutionary changes build steam and spread throughout a population, cascading through a species, funneling it through particular environments, over pressure humps, and around the threat of extinction, along the way turning it into another species entirely.

Awareness of our impending death seems to be an artifact of language's reaching a certain stage of complexity. Now we are coming to another realization about language and our species that may be the seed of an equally profound idea. We have believed for a long, long time that language is a monolithic thing. But all the evidence reported in this book argues that it is not. The bottom line is that language is not how we intuitively think of it. As Terrence Deacon says, language is not language in the way the Lego is Lego. Lego is Lego all the way through, but language, which we experience as an integrated whole, is instead a bitsy pile of stuff, some parts ancient and others less old.[11]

It turns out that the same can be said about us—neuroscience indicates that individuals are no more unitary or whole than language is. We think of ourselves as single creatures, but as individuals and as a species we are assemblages of traits, features, and experiences, and these all shift in relative importance in different contexts. Certainly, language is fundamental to our identity. It shapes who we are in ways that are irreversible, and there is no going back to who you were before you were taught to speak. But if we weren't taught, we would never speak. As evolution works upon us, it may choose to elaborate parts of ourselves that we don't really see or elements of behavior that we don't regard as separable from the rest of us. In this way, our descendants might become unrecognizable to us. There is only so much destiny in our genome—life arises when DNA and the world wind together, and that's not in our control.

Think back now to the worldwide language web. Imagine all the language networks, parent to child, that extend from the present back through time. It's small wonder that humans dream in myth and in art about other worlds, because we all have the experience of inhabiting one world and, as we are taught language, of walking through a door into another. Even physicists are obsessed with the idea of a multiverse. But we already live in one.

Epilogue: The babies of Galápagos

Because the revolution in language evolution is so recent, one of the most important messages of this book is the very basic idea that investigating the evolution of language is a good and worthwhile pursuit. Indeed, it's not possible to fully understand language if you don't take evolution into account—either you must begin with evolution or you must make room for it.

I have tried to draw attention to the commonalities among language evolution researchers, and it may turn out that many of the researchers who disagree have more in common than currently seems to be the case. At the moment, most scientists are not particularly concerned with tying the stages they consider important to language evolution to the chronology of our evolutionary history. But as the research becomes more elaborated, what now look like conflicting theories of linguistic expansion may end up as different phases in the same evolutionary account.

This epilogue is devoted to the differences among researchers. Part of the glory of the language evolution debate, as with all the other big, messy debates, like that about the relationship between mind and brain, is just how many highly trained and really smart people disagree completely with one another.

It used to be that words like "innateness" and "uniqueness" were sufficient to pinpoint the distinctions between particular scientists or schools of thought. Until recently, a great divide separated those who believe there is some kind of computational mechanism at work in the generation of language and those who think it can be explained only by general principles. But even now, these two positions are becoming more difficult to distinguish. No one serious has taken a stance at an extreme end of the continuum, and each side makes concessions to the other.

Everyone would agree that our biological endowment and the way that our individual lives unfold cannot be fully disassociated. In fact, we need an easy word to describe what we actually do have: a unity of nature and nurture. Geneticists talk about the phenome, the inextricable mesh of the individual's genome and the environment that selects and deselects the way the genome gets expressed. Probably the best word for our purposes is just "life."

But even without extreme arguments, examining the role of biological endowment and the environment in language learning remains one of the best ways to identify differences between scholars who differ in the relative weight they assign to each. Chomsky once likened the emergence of language to the growth of limbs, implying that language is something that inexorably projects out of the individual without effort or conscious intervention. Other researchers like Philip Lieberman cite cases like Genie, the little girl who was not spoken to as a child and never developed language normally. He maintains that children must be exposed to language in order to acquire it fully.

I asked the key researchers interviewed for this book to answer the following question (some declined to participate): *If we shipwrecked a boatload of babies on the Galápagos Islands—assuming they had all the food, water, and shelter they needed to thrive—would they produce language in any form when they grew up? And if they did, how many individuals would you need for it to take off, what form might it take, and how would it change over the generations?*

Michael Arbib: The closest data that we have on this topic is that of Nicaraguan Sign Language. Here, a group of deaf children, brought together in a school for the deaf in Managua, Nicaragua, spontaneously developed a full human signed language over three "generations" (where a generation was not a biological generation but rather a cohort of children admitted over a ten-year period). Each cohort seemed to plateau in its capability, so that the signing of the first generation was more like pantomime and less like conventionalized sign language. But with each generation, the repertoire of conventionalized signs and the expressivity with which they combined increased greatly.

Let's leave aside the fact that babies given food, water, and shelter but without caregivers are unlikely to survive. Assuming they did survive as an interacting group, the data on Nicaraguan Sign Language might suggest that if a group of babies were raised in isolation from humans with language (the Galápagos Islands really don't qualify), then in three generations (this time, biological generations) some critical mass of children—let's say thirty or so—would develop language. And presumably, since these children are not deaf, one might well expect the resulting language to combine vocal and manual gestures, as does normal human discourse.

However, I doubt very much that this would happen. I believe that the brain of *Homo sapiens* was biologically ready for language perhaps 200,000 years ago, but if increased complexity of artifacts like art and burial customs correlate with language of some subtlety, then human languages as we know them arose at most 50,000 to 90,000 years ago.

One may either respond by rejecting this idea that it took human brains 100,000 years or more to invent language as we know it or suggest that the Nicaraguan deaf children had an advantage that early humans lacked. I adopt the latter view. And what is that advantage? I claim that it is the knowledge that things can be freely named, and the knowledge that languages do exist. Certainly, the Nicaraguan children could not hear, but they could see the lip movements that indicated that their families could communicate their needs and requests. In addition, they lived in a world of many distinctive objects, both natural and artificial, and could see that something more subtle than pointing could be used to show which object was required. Moreover, some had at least basic knowledge of Spanish and had both seen and performed a variety of co-speech gestures. They thus would be motivated to try to convey something of their needs, or share their interest by pantomime and the development of increasingly conventionalized gestures. Intriguingly, Ann Senghas (an expert on Nicaraguan Sign Language) has told me that the second generation even went to Spanish dictionaries in search of words for which they needed to develop hand signs.

For us, as modern humans, it seems inconceivable that the very idea of language is something that has to be invented. Yet, to take a related example, we know that writing was invented only some five thousand years ago. Yet once one has the idea of phonetic writing, it is a straightforward exercise to invent a writing system—as has been demonstrated by many Christian missionaries who wanted to bring literacy and the Bible to a people who had language but no writing.

In view of all this, I doubt very much that a few children on a desert island would develop much beyond a rudimentary communication system of a few vocal and manual gestures and some conventionalized pantomime unless they had hundreds of generations in which to create culture and the means to discuss it. But they would have the brains to support such inventions, whereas other creatures would not.

Paul Bloom: The answer is: yes, two. This is more than guesswork because a variant of this situation has already occurred, more than once. This is when children grow up without being exposed to a language model, such as deaf children who are raised by adults who don't use sign language. Such children will sometimes create a rudimentary language, complete with words and some sort of morphosyntax. Over generations, the vocabulary will grow, and the syntax and morphology will become more complex.

Wolfgang Enard: Yes, if there are more than two.

Tecumseh Fitch: Yes. You'd need a village worth. They start out with something very basic in the first generation. Then they'd develop a pidgin in the second generation. In the third they'd have a creole, and by the fourth they'd have a fully stable language.

Marc Hauser: Since the language faculty requires input of some kind in order to be expressed as an externalized or e-language in Chomsky's sense, there would be highly structured internal thought but there would be no expressed language and no communication, as there would be no one to communicate with.

Ray Jackendoff: Yes. Only if there were about thirty of them to begin with.

Simon Kirby: There are two different sources of evidence that we can use to get a handle on these Galápagos children. Firstly, because computational agents aren't yet complex enough to have acquired any rights as experimental subjects, we are free to re-create this Galápagos in computer simulation. This very scenario has been the subject of a good deal of research, particularly over the past decade. One of the problems with this approach is that it can tell us only what might happen, given particular assumptions about how children's brains work, and that in itself is a difficult research issue. However, we *can* learn from simulation exactly how little, or how much, of language is required to be pre-wired before a language would emerge in the Galápagos population, and what factors other than individual psychology are important.

A second source of evidence comes from studies of spontaneous language emergence in real populations. Of particular interest are the indigenous sign languages that have evolved in populations of deaf children who lack a preexisting shared system for communication. The most celebrated case is the language that emerged in Nicaragua around the time of the Sandinista revolution when a large population of deaf children were brought together for the first time. However, a fascinating meta-study conducted by the anthropologist Sonia Ragir has shown that this kind of language emergence is not guaranteed to occur in all populations of deaf children who lack a common language. Just as has been suggested by the computational models, certain features of the population (and its dynamics over time) appear to be critical for a novel language to emerge.

I would bet that the emergence of what we'd consider a "full" human language in the Galápagos scenario is equally not guaranteed. What would make it more likely is if the population of children was large, if further boats of children arrived at regular intervals (say once or twice a year), and if every member of the community was engaged socially with the group in a way that made linguistic communication relevant. I would expect that language emergence would be gradual, with later arrivals and younger arrivals using the language in increasingly abstract ways and with increasing amounts of what we'd consider linguistic structure. I would expect that for a good number of generations a visiting linguist would be able to tell straightaway that the Galápagos language was unusual. Whilst it would be likely to "obey" some of the key universal principles of language organization—particularly ones that have their explanation in language processing and use—it would lack many of the morphological irregularities and paradigmatic quirks that are common in "normal" languages (i.e., languages with a history). If I were to go out on a limb, I'd say it is at least possible that the language would initially lack some of the features that we take for granted, like potentially unlimited embedding (my brother's son's friend's mother's fishing net), and perhaps some basic grammatical categories. In addition, we can be almost certain that for a very long time it would lack an extended numeral system or more than a handful of basic color terms.

I think the Galápagos experiment might also show us that much of the complex structure of human language doesn't really give us immediate payoffs in terms of any enormously increased chance of survival. I would predict that the vast majority of the survival needs of the nascent community would be served by a much more primitive protolanguage. If a structured, complex language were to emerge, it might not be of any immediate survival benefit (at least as a result of communicative efficacy); rather, structure and regularity would appear purely because of the adaptive dynamics arising from the way the language itself is transmitted from individual to individual. Only much later, as more and more complex cultural artifacts appeared on the Galápagos, would having a complex human language come into its own as a way of transmitting cultural information from individual to individual.

If this is right, it suggests a problem for explanations of language structure that rely on natural selection pressures arising from communicative needs. Why then would we have the ability to acquire complex language at all? Of course, this is the question we all want to answer. Certainly, we already know that signaling systems with fairly complex structure have evolved a number of times in nature without providing any obvious communicative benefit. Perhaps the visiting linguists would also learn a lot from listening to the songs of the birds on the island, or to any whale song they might hear on their voyage there.

Chris Knight: The key innate feature of human cognition—the one most relevant to the emergence of language—is in my view the capacity for joint attention and egocentric perspective reversal. We humans possess an inborn capacity to correlate our perspectives on the world, viewing ourselves from one another's standpoint. As well as being cognitive in the narrow sense, this faculty has, simultaneously, moral relevance. If I choose to have a violent tantrum, I must temporarily shut down my moral self-awareness. I need this kind of awareness only if I am trying to tune my behavior to social requirements.

For this reason, I imagine the boatload of babies would spontaneously produce some kind of language. But if I think this, it's because I assume the population would comprise females and males in about equal proportions, and because I assume potentially violent conflicts over sex would be sorted out as these individuals reached puberty. A potentially aggressive, sexually violent male, for example, would soon meet collective opposition. If this was effective, it would force him to view his behavior from the standpoint of others, modifying that behavior accordingly. The idea that language as we know it could emerge wholly autonomously, in isolation from any kind of institutional structure—any kind of self-organized moral regulatory framework—is gravely mistaken.

Like the use of paper money, linguistic communication depends entirely on trust. If the social and sexual dynamics on the Galápagos Islands obstructed the emergence of sufficient mutual trust, then the emergence of a shared public language in the shipwrecked population would be severely threatened and obstructed. As we all know, chimpanzees, bonobos especially, have considerable innate potential for symbolic communication. Why is such potential not drawn upon in the wild? The reasons are political as much as cognitive. Chimpanzees don't hold one another to collectively agreed standards of public behavior. Although as individuals they value cooperation and sociability, it soon becomes clear that a kind of internal civil war is the default state in relations between sexually mature adults. Sometimes this civil war is latent; sometimes it explodes into the open. Where public trust is not the default state, individuals have no choice but to fall back on emotionally persuasive, hard-to-fake gesture-calls. This applies even to humans. One theoretically possible outcome in the Galápagos would be something like the scenario depicted by William Golding in his terrifying novel *Lord of the Flies*. Under violent and inhuman conditions, I imagine that this shipwrecked population would engage in a lot of screaming, crying, and so on, but not a lot of quiet, rational conversation.

Philip Lieberman: No.

Gary Lupyan: The emergence of language on the island is by no means certain even with a fairly large number of individuals. I think the emergence of modern language is as much a cultural as a biological phenomenon.

It's helpful to think of the emergence of language in terms of other cultural achievements of our species such as the emergence of writing. Biologically modern humans are obviously capable of reading and writing, but if you put a bunch of already-speaking babies on an island, what's the chance that writing would emerge within their lifetimes? Well, it happened at some point, so it's not zero. But the chance depends on factors like their motivations, their culture, their technology, and so on. Hunter-gatherers have little need for writing. But given a culture with money, farmers, and landowners, and you can foresee a need to keep permanent records.

Like writing, the emergence of language depends on motivational and environmental factors. What those are is a bit of a mystery. Group size is an obvious one; the need to organize and cooperate is a likely factor. If food is plentiful on this island and individuals can get by in small groups, the emergence of language is less likely.

If language emerged, I think it would change radically over the first few generations, demonstrating the kinds of changes languages undergo as they move from pidgin to creole. But language change is not just something that happens with time—change responds to pressures of the environment and the society. If the people on the island needed ways to talk about time in a precise way, we could expect a complex system of tenses to emerge. If the culture created by the children is very hierarchical, a grammaticized system of honorifics may emerge in the language.

I believe the idea of language as an instinct is wrong. People who hold this view are so impressed with children's proficiency in acquiring language that they take for granted that modern children are born into a linguistic environment. The hard part—"inventing" a language—has already been done for them. From the day they are born, children interact with people whose minds and behaviors have been shaped by language.

In the case of the Nicaraguan deaf children, though they couldn't understand the words spoken by their hearing parents, the children were interacting with people who did possess full-blown languages. In creating their own sign language, these kids were not copying the languages of the people around them, but this process of creation was in no small part launched by the linguistic environment in which they found themselves.

Heidi Lyn: No to "language." I believe they would develop a protolanguage communication system, but if there was no need for cooperation and no culture to model communication for them, I don't think they would develop full-blown language. I think you would need at least ten individuals for any kind of real communication to take place. I still think this protolanguage would be more developed than what we've seen in other species, but with only a few individuals I don't think there would be any need even for that.

Gary Marcus: Yes. You'd need two. Language builds on our cognitive capacities to reason about the goals and intentions of other people, on our desire to imitate, our desire to communicate, and our twin capacities for using convention to name things and sequence to indicate differences between differing possibilities. Two children who started a language afresh might have simpler language—a smaller set of sounds, nouns and verbs, but not adjectives or adverbs, little (perhaps nothing) in the way of inflectional morphology. Languages certainly become richer over time, and a language that started from scratch would in its first generation be limited. But put two human children together on a desert island, and I bet they would develop a system that looks a lot more like language than anything found in any other species in the natural world.

Irene Pepperberg: Will the babies develop some kind of communication system? Very likely, as some innate predisposition to bond, gain the attention of others, interact, and so on seems to exist. Will this communication system initially be anything as complex and sophisticated as "regular" human language? Unlikely . . . Would it increase in complexity over generations? Tricky question . . . That depends on how much information will need to be communicated.

Steven Pinker: I'd guess that the children would create a simple but fluent language, perhaps with a mixture of signs and speech, somewhere between a creole and Nicaraguan Sign Language.

Luc Steels: Yes. They would develop language. It's a social institution. There has to be a joint problem to solve—food or navigation—and then it would happen within one generation. You learn a language by building it. You invent it on your own.

ACKNOWLEDGMENTS

For their important work and their willingness to share it, thank you, Michael Arbib, Kate Arnold, Derek Bickerton, Paul Bloom, Josep Call, Dorothy Cheney, Noam Chomsky, Morten Christensen, Terrence Deacon, Frans de Waal, Frederic Dick, Laurance Doyle, Wolfgang Enard, Dan Everett, Ramon Ferrer i Cancho, W. Tecumseh Fitch, Timothy Gentner, Susan Goldin-Meadow, Peter Gordon, Melissa Groo, Marc Hauser, William Hopkins, Gavin Hunt, Jim Hurford, Ray Jackendoff, Alex Kacelnik, Ben Kenward, Simon Kirby, Chris Knight, Leah Krubitzer, David Leaven, Philip Lieberman, Heidi Lyn, Vittorio Loreto, Gary Lupyan, Gary Marcus, Lori Marino, John McWhorter, Svante Pääbo, Katy Payne, Irene Pepperberg, Massimo Piatelli-Palmarini, Steven Pinker, Diana Reiss, Duane Rumbaugh, Robert Seyfarth, Sue Savage-Rumbaugh, Brad Schlaggar, David Schwartz, Katie Slocombe, Luc Steels, Mike Tomasello, Carel van Schaik, Faraneh Vargha-Khadem, Janette Wallis, and Klaus Zuberbühler.

My reporting on brain surgery at Johns Hopkins was done while on assignment for *The New Yorker*. My sincere appreciation to the Nissley family and the doctors and researchers at Johns Hopkins, and also to Kim Hoppe. Many thanks as well to Leo Carey, Raffi Khatchadourian, and Henry Finder. Some of the reporting and writing on language and music was carried out for the *Boston Globe* "Ideas" section. Thanks to Alex Star.

Teachers, editors, writers, linguists, and friends who have generously given practical help and imparted wisdom about their craft and interests include Sidney Allen, Melissa Anderson, Michael Anderson, Jayne Ashley, Elaine Blair, Elizabeth Brown, Katherine Chetkovich, Caleb Crain, Joe Doebele, Nick Evans, Suzy Hansen, Emily Loose, Ellen Maguire, Peter H. Matthews, Peter Meyers, Lisa-Jane Moody, Francis Nolan, Rachel Nordlinger,

Cliona O'Gallchoir, Amanda Schaffer, Sue Shapiro, Peter Terzian, Lawrence Weschler, and Jenny Wiggins. Thanks and g'day to Helby, Tubs, Son, and Woz.

For their invaluable advice, assistance, and cheer, I am indebted to my monthly authors' group. It includes Gary Bass, Susan Devenyi, Elizabeth DeVita-Raeburn, Abby Ellin, Sheri Fink, Katie Orenstein, Pamela Paul, Heidi Postlewait, Alissa Quart, Paul Raeburn, Deborah Siegel, Rebecca Skloot, Stacy Sullivan, Harriet Washington, and Tom Zoellner. Many, many thanks to the experts who have been our guests. In the same vein, I have great admiration for Kamy Wicoff and Nancy K. Miller for their wonderful salon.

Much love and thanks to Deveka Leibovitz, Jessika Milesi, and Camille Cadougan for doing the most important job in the world. Along these important lines, I am profoundly grateful for the camaraderie of Jessica Alger, Anne Baker, Jessica Bauman, Holly Kilpatrick, Robbyn Kistler, Rose Ricci Mullen, Brigid Nelson, and Kate Porterfield. I would also like to express my humble appreciation to Eva Sosnowska.

Thanks to Oren Arnon, Rachel Bingman, Gerry Cole, Ben Grasso, Emily Guzzardi, Jesse Hewit, Kevin Johnston, Amanda Key, Ali Narimani, Bridget Scruggs, Josh Sidis, Alma Steingart, Rachel Stogel, Peter Twickler, and Greg Wolf for solving the problem of consciousness, or at least my problem with it. Writing this book wouldn't have been half as much fun without the Tea Lounge. Thank you, Libba Bray and Ben Jones for your delicious companionship.

By meeting almost every week for a couple of excellent years, Marci Alboher, Sarah Milstein, and I bootstrapped ourselves from one place to another altogether. We also ate a lot of great food. Cheers, goils.

Nothing happens without the love and support of Desmond Kenneally, Josephine Kenneally, Hugh Kenneally, Katherine Milesi, Angela Kenneally, and Shelagh Lloyd. I appreciate them all so much, and I thank them for Janet McAllister, Sam Trinder, Madeleine Kenneally, Steve Milesi, Jessika Milesi, Justin Milesi, Mick Jukes, Simon Lloyd, and Allegra Lloyd.

All writers need serenity to accept the things they cannot change, courage to change the things they can, and an agent to tell them the difference. For this, among his many talents, I am so lucky to have Jay Mandel at the William Morris Agency. Rick Kot at Viking gave me all the space I needed to write this book, and then he applied his immense skills to every line. He is not just a patient and engaged editor; he is the kind of person who makes you want to write really well. Thank you, Greg Mollica and Nicholas Blechman, for creating a beautiful cover, and Francesca Belanger, for a wonderful book design. It's been an education and a boon to receive the ministrations of Bruce Giffords and Ingrid Sterner, and I appreciate the help of Talia Cohen, Liza Monroy, Charlotte Wasserstein, Laura Tisdel, and Sabila Khan.

Thanks to Lorin Stein, whose delightful friendship and expert guidance helped kick this book off, and to the brilliant Annie Murphy Paul, who had the last word on the first three.

Finally, a perpetual motion machine's worth of thanks to Christopher Baldwin, and also to Nathaniel and Fineas, who give evolution all the meaning it needs.

NOTES

Many interviews were conducted in the course of researching this book, taking place over a period of five years. Some were face-to-face, in offices, in laboratories, and at conferences. Some were carried out on the phone and others by e-mail. Often, all three methods were used with one researcher. If the source of a quote is not apparent from an endnote reference or from the text itself, the citation originated in a personal interview.

Prelude

1. Terrence Deacon discusses the way the virtuality of language and the physicality of the world intersect in *The Symbolic Species*.

Introduction

1. Morten Christiansen and Simon Kirby ask if language evolution is the hardest problem in science in the introduction to M. H. Christiansen, S. Kirby, *Language Evolution*.

2. Thanks to Simon Kirby for being the first to track the number of papers and for pointing in the direction of http://www.isrl.uiuc.edu/amag/lang ev/.

3. C. Darwin, *On the Origin of Species*, 162.

4. Peter Matthews was my Ph.D. supervisor.

I. Language Is Not a Thing

Prologue

1. J.-J. Rousseau, V. Gourevitch, *The First and Second Discourses Together with the Replies to Critics and Essay on the Origin of Languages*. In order to get around the religious establishment, Rousseau posited a first language that was given to Adam by God. Somehow we lost it along the way. According to Jean Aitchison, this creation-and-destruction myth allowed Rousseau to talk about evolution. See J. Aitchison, *The Seeds of Speech*, 6.

2. E. Mayr, *What Evolution Is*, vi.

3. C. Darwin, *The Descent of Man*, 89.

4. Ibid., 88.

5. Ibid., 189.

6. Ibid., 87.

7. Ibid., 92.

8. Ibid., 89.

9. He also noted that just like animal species, once a language becomes extinct, it never returns.

Similarly, no language or species ever appears in two different birthplaces.

10. V. Volterra et al., "Gesture and the Emergence and Development of Language."

11. G. W. Hewes. "Primate Communication and the Gestural Origin of Language," 6.

12. Other conferences discussed the topic within a broader context, such as another NATO summer institute in Italy organized by Philip Lieberman and two conferences sponsored by the Wenner-Gren Foundation for Anthropological Research that took place in 1986 and 1990.

Chapter 1. Noam Chomsky

1. R. Blackburn, O. Kamm, "For and Against Chomsky."

2. Of course, field linguists are still hard at work today, racing to record the world's languages and describing the relationships among them. Linguistic fieldwork requires not only extreme diligence but a mix of intellectual brilliance with a certain pragmatism.

3. It didn't get published for another twenty years.

4. R. A. Harris, *The Linguistics Wars,* 38.

5. Roman Jakobson Collection, MIT Archives.

6. R. A. Harris, *The Linguistic Wars,* 39.

7. Ibid.

8. Chomsky did not invent the term or the basic idea, but his version of it animated the field.

9. R. A. Harris, *The Linguistics Wars,* 39. Over time, the implication shifted; more and more there was a sense that grammars were psychologically real, and that phrase structures rules existed inside the heads of speakers.

10. T. Grandin, C. Johnson, *Animals in Translation,* 9–10.

11. N. Chomsky, "A Review of B. F. Skinner's Verbal Behavior," 26–58.

12. This was with reference only to humans. Chomskyan theory, if not Chomsky himself, continued to contribute to the view that animals are unthinking machines.

13. Randy Allen Harris brings together a great deal of such evidence in chapter 3 of *The Linguistics Wars.*

14. Ibid., 67.

15. Howard Maclay, "Overview," 163.

16. R. Jackendoff, *Foundations of Language,* xi.

17. D. C. Dennett, *Darwin's Dangerous Idea,* 385.

18. C. Knight, "Decoding Chomsky," 581.

19. Chomsky has been called Copernican, Newtonian, Einsteinian, Planck-like. For both its significance and its revolutionary character, his work has been compared to that of Spinoza, Pierce, Wittgenstein, Darwin, and Freud. He is an angel, a God, an enfant terrible. Supporters criticize him with the inevitable caveat "Noam Chomsky is one of the half-dozen great geniuses of the twentieth century." In 1990 Derek Bickerton wrote in *Language and Species*: "Most of what we know about language has been learned in the last three decades." Four years ago in *Science,* David Premack wrote that "linguistics is the science of sentences"—an entirely Chomskyan

way of looking at it. Alternately, Chomsky has been described as satanic, the Enemy, a crank, an embarrassment. Generative linguistics has been called a cult; generative linguists have been described as "born again." An extensive literature analyzes everything Chomsky has wrong, from the misinterpretations that he made of Skinner to all the ways he ignores contrary evidence. Today, people writing in Internet mailing lists work themselves into apoplectic rages about statements he allegedly made twenty years ago. A recent article said he has led linguistics down a blind alley.

20. R. A. Harris, *The Linguistics Wars*, 73.

21. Ibid., 67–68.

22. George Lakoff, Haj Ross, James McCawley, Jerrold Katz, Paul Postal.

23. R. A. Harris, *The Linguistics Wars*, 201.

24. P. H. Matthews, *Grammatical Theory in the United States from Bloomfield to Chomsky*, 250.

25. R. A. Harris, *The Linguistics Wars*, 157.

26. One phenomenon for which Chomsky and his early colleagues in transformational grammar have been held responsible is the sour way researchers in the field used to deal with disagreement. In 1998 James McCloskey wrote that while other types of linguists manage to get along, "syntacticians seem to thrive on a more robust diet of anger, polemic and personal abuse." From R. A. Harris, *The Linguistics Wars*, 80.

27. N. Chomsky, *Reflections on Language*, 4.

28. P. H. Matthews, *Grammatical*

Theory in the United States from Bloomfield to Chomsky, 237.

29. Many histories of twentieth-century linguistics are concerned with misapprehension. Most of them set out to correct the erroneous impressions and mythmaking of the other tomes. No one seems to be telling a straight story, and everyone addresses the fact that other writers don't have it right. Some books propose candidate publications as the moment when things really changed, and others argue that, no, that wasn't the first publication to change everything, it was the second.

30. N. Chomsky, *Language and Mind*, 97–98.

31. Ibid.

32. N. Chomsky, "Discussion of Putnam's Comments," 321.

33. Elsewhere, Chomsky suggested that the question of language evolution was of no more interest than the evolution of the heart.

34. G. W. Hewes, "Primate Communication and the Gestural Origin of Language," 6.

Chapter 2. Sue Savage-Rumbaugh

1. Goodall's research and that of scholars who followed her lead resulted in the installment of primatology centers all over the United States in the 1940s. Most of our knowledge about our closest relatives comes from the fact that one individual, like Goodall, has hunched down in the wild or at a primate facility and spent many hours watching them.

2. R. M. Sapolsky, *A Primate's Memoir*.

3. E. S. Savage-Rumbaugh, S. Shanker, and T. J. Taylor, *Apes, Language, and the Human Mind,* 25.

4. Gardner, M. "Monkey Business."

5. E. S. Savage-Rumbaugh, S. Shanker, and T. J. Taylor, *Apes, Language, and the Human Mind,* 207.

6. D. M. Rumbaugh, W. D. Hopkins, D. A. Washburn, and E. S. Savage-Rumbaugh, "Comparative Perspectives of Brain, Cognition and Language," 45.

Chapter 3. Steven Pinker and Paul Bloom

1. S. J. Gould, "Evolution: The Pleasures of Pluralism."

2. S. J. Gould, *Full House,* 216. Richard Dawkins, Charles Simonyi Professor at the Oxford University Museum of Natural History and author of many books on evolution, including *The Selfish Gene* (New York: Oxford University Press, 2006), has stressed in his work the ways that evolution is nonrandom. For instance, mutation may be random, but selection is not. Rather, the animals that are the best at what they do survive—there is nothing random about this. Dawkins and Gould vigorously debated this, and other topics, over many years.

3. The Internet has made technical reports obsolete.

4. S. Pinker, P. Bloom, "Natural Language and Natural Selection," 710.

5. Ibid., 708.

6. But see John McWhorter's comment on p. 287.

7. C. Darwin, *On the Origin of Species,* 186.

8. Said Pinker: "The talk was given at the MIT Center for Cognitive Science seminar series (I don't know if the series itself had an official name, but it was well known around the Boston area). The series ran from the mid-1970s to the mid-1990s. A paper was circulated in advance, and the authors defended it and engaged with one or two commentators. It was assumed that the audience had read the paper; the authors presented a brief summary, then gave the stage to the commentators, then replied, and a discussion followed. It was a terrific format, and there were often fireworks."

9. S. Pinker, P. Bloom, "Natural Language and Natural Selection," 710.

Chapter 4. Philip Lieberman

1. P. Lieberman, *The Biology and Evolution of Language.*

2. C. Darwin, *On the Origin of Species,* 190.

3. P. Lieberman, *Human Language and Our Reptilian Brain,* 23.

4. Not only did this technique *not* help the fever, explained Lieberman, but wounded soldiers at Waterloo were more likely to survive if they were *not* treated by surgeons.

5. P. Lieberman, *Human Language and Our Reptilian Brain,* 166. For all their differences, Lieberman is not unlike Chomsky. Both men have a reputation for being provocative and intimidating. When I speak to Lieberman, he is as frank in his assessment of his opponents' intelligence as Chomsky is. "He's an idiot!" he says of one researcher. "That paper is crazy!" he says of another's publication. Chomskyan linguistics, according to Lieberman, is based on

"infant school mathematics." Still, like most of the researchers I speak to, he is ultimately democratic. At the end of a conversation in which he laid waste to the ideas of one academic, he added with complete sincerity, "Of course, you should go talk to him."

6. In a 1990 e-mail to Chomsky, Pinker wrote that he thought the subject was one about which reasonable people could disagree.

II. If You Have Human Language . . .

1. There is, apparently, a firm bottom end of 19,599 genes. In a similar example of anthropocentrism, a special type of brain cell, known as a spindle cell, has been long thought to be specific to humans and great apes. In 2006 it was announced that whales also have spindle cells.

2. Neuroscientists don't get squeamish, said de Waal, because neuroscientists know that the emotional centers we have are exactly the same as those of other animals. They can be studied in animals, even rats. If you look at the dualisms that we live in, he said, such as the dualism between human and animal, mind and body, nature and culture, you see that the dualism between mind and body has completely fallen apart under the influence of neuroscience.

3. If crocodiles had human language, there'd be no lurking alone along the riverbank or floating loglike and dangerous on the water's surface. Instead, they would be constantly communicating, maybe comparing

notes on death rolls and indigestion. Instead of birthing fifty babies and indifferently letting them loose on the world, they'd be talking to them in exaggerated tones, cherishing their every sweet croak, and construing their flailing little gestures and sounds as meaningful. They would, in short, be human, not crocodilian.

4. In *The Ancestor's Tale,* Richard Dawkins follows the trail of the human species back through time, highlighting where the path of our ancestors links up with the paths of other species' ancestors.

5. Other tools enjoyed by orangutans include cleaning implements. Lyn used to give an orangutan a bucket and a mop, and it would clean its cage to entertain itself. She also gave the ape a long stretch of rope, and it would spend hours tying and untying knots. Orangutans are not just good tool users; they are very good at deception. They have been known quietly to conceal metal in their mouths, and once their keepers have gone for the day, to use it to open their cage doors. (Usually you can tell if a chimpanzee is trying to conceal something—it just can't hide its excitement.) One apocryphal orangutan story is about an ape that would hide metal in his mouth, every night let himself out of his cage, get a mop and a bucket, and then let himself back in. Each morning his keepers would find him cleaning. The joke, of course, is that they'd let him finish cleaning before they took the contraband metal away. There is a great deal more documentation for chimpanzee tool use in the wild and in

experimental situations. It's been suggested that the main difference underlying tool use in chimpanzees and orangutans is one of temperament.

Chapter 5. You have something to talk about

1. Animal tool use was debated in Darwin's time. In *The Descent of Man,* he notes (p. 83) the claim that tool use is a uniquely human endeavor and provides many counterexamples: "It has often been said that no animal uses any tool; but the chimpanzee in a state of nature cracks a native fruit, somewhat like a walnut, with a stone . . . I have myself seen a young orang put a stick into a crevice, slip his hand to the other end, and use it in the proper manner as a lever. The tamed elephants in India are well known to break off branches of trees and use them to drive away the flies; and this same act has been observed in an elephant in a state of nature . . . Brehm states, on the authority of the well-known traveller Schimper, that in Abyssinia when the baboons belonging to one species (*C. gelada*) descend in troops from the mountains to plunder the fields, they sometimes encounter troops of another species (*C. hamadryas*), and then a fight ensues. The Geladas roll down great stones, which the Hamadryas try to avoid, and then both species, making a great uproar, rush furiously against each other. Brehm, when accompanying the Duke of Coburg-Gotha, aided in an attack with fire-arms on a troop of baboons in the pass of Mensa in Abyssinia. The baboons in return rolled so many stones down the mountain, some as large as a man's head, that the attackers had to beat a hasty retreat; and the pass was actually closed for a time against the caravan. It deserves notice that these baboons thus acted in concert."

2. E. D. Jarvis et al., "Avian Brains and a New Understanding of Vertebrate Brain Evolution."

3. The "renaming" of the bird brain was an unusual event. Scientists generally get on with their work using the language already at their disposal. (See the writing of Richard Lewontin and Terrence Deacon for particular sensitivity to the effect of metaphors on science. Also, recall Philip Lieberman's explanation of the effect of the brain-as-computer and other metaphors in chapter 4.) One implication is that we will now be able to draw a more direct comparison between birds and mammals. On the publication of "Avian Brains and a New Understanding of Vertebrate Brain Evolution," one scientist suggested that the bird will become the new laboratory rat. This may be overenthusiastic. Rats, after all, are phylogenetically closer to humans.

4. Many explanations for mirror self-recognition after the first Gallup experiments were couched in terms of the primate body or brain. One theory had it that self-awareness somehow evolved from handedness, arising as we swung, one hand after the other, through the trees. Another explanation said that self-recognition was associated with our unique frontal lobes. Of course, dolphins don't have hands, and they have a very different brain structure.

5. K. Wynn, "Addition and

Subtraction by Human Infants." Note that some researchers do not believe Wynn's experiments show sensitivity to number but to mass/contour instead.

6. E. M. Brannon, H. S. Terrace, "Representation of the Numerosities 1–9 by Rhesus Macaques (*Macaca mulatta*)." Terrace is the scientist who trained Nim Chimpsky.

7. S. Dehaene et al., "Core Knowledge of Geometry in an Amazonian Indigene Group."

8. K. E. Jordan, E. M. Brannon, "The Multisensory Representation of Number in Infancy."

9. If we hope to judge fairly an animal's ability to think, researchers must enter their world as much as possible. Payne exemplifies this more than most. Twenty years ago, she pursued a hunch and discovered that much elephant communication takes place over long distances, well below the level of human hearing. Until Payne's finding, no one knew that we had been seeing and hearing only half of the elephant's world. What will the mental world look like when we no longer insist on measuring it in terms of human intelligence? Payne believes the traditional ordering of nature into a pyramid with *Homo sapiens* at the top will change; *"sapiens,"* said Payne, "is a hilarious term."

10. It's not that animals are mute thinkers. They are fully expressive for *their* purposes. However, our form of communication has a profoundly ramifying effect on our thought.

11. S. Pinker, R. Jackendoff, "The Faculty of Language: What's Special About It?" 6.

12. N. J. Mulcahy, J. Call, "Apes Save Tools for Future Use."

13. S. Pinker, R. Jackendoff, "The Faculty of Language: What's Special About It?" 5–7.

14. Pinker and Jackendoff distinguish between numbers in this regard and those that can be subitized.

15. A. L. Gilbert et al., "Whorf Hypothesis Is Supported in the Right Visual Field but Not the Left."

16. P. Gordon, "Numerical Cognition Without Words."

Chapter 6. You have words

1. R. M. Seyfarth, D. L. Cheney, P. Marler, "Monkey Responses to Three Different Alarm Calls."

2. Not everyone agrees with the signature whistle interpretation. See B. McCowan, D. Reiss, "The Fallacy of 'Signature Whistles' in Bottlenose Dolphins."

3. V. M. Janik, L. S. Sayigh, R. S. Wells, "Signature Whistle Shape Conveys Identity Information to Bottlenose Dolphins."

4. Of course, for all three, the vocalizations occur in completely different social and biophysical contexts. Human speech is transmitted through the air. Dolphins make sound underwater, and elephants communicate across a wide pitch range (compared to ours) both through the atmosphere and the ground.

5. The exception to this rule is onomatopoeia, where the sound of a word evokes the sound or action it describes—e.g., hiss, tinkle, buzz, hum. These words are rare, and the closeness of their sound-meaning connection is fairly subjective.

6. S. Pinker, R. Jackendoff, "The Faculty of Language: What's Special About It?" 12.

Chapter 7. You have gestures

1. Baboons have a rich repertoire of gestures in addition to the muzzle wipe. The adult males exchange complicated greetings, where they make particular facial expressions, assume certain postures, embrace each other, and briefly handle one another's genitals—kind of like a handshake but with the most vulnerable part of the body.

2. Wallis was introduced by Josep Call, a highly experienced researcher, and afterward, Call showed a video of chimpanzees in the wild, mentioning that he hadn't noticed any muzzle-wipe behavior in the animals. On the spot, Wallis got him to replay some frames of his video. She pointed out at least five examples of a movement that looked like a muzzle wipe. Call was visibly startled to see the gesture, then he laughed and turned to the audience: he recounted yet another time that Wallis was told by a primatologist that he had never seen the gesture she was talking about. Then, too, Wallis got the primatologist to replay some of his own footage, and she pointed out what he had missed. Some make the case that we've been observing chimpanzees and other animals for so long now— fifty years—we are not going to find anything that would surprise us. And yet until Wallis showed her videos of the baboons' muzzle wipe, it could have been said that there was no evidence for this gesture—despite the fact that it exists and is ubiquitous.

Wallis was the only one to see it and take it seriously, and her experience shows how easy it is for experts to miss what is right before their eyes.

Ideally, science would be based on observations of all reality, but it is not like this, and animal science is even less so. Instead, the picture is blotchy. Each researcher who announces findings about animal behavior has made choices along the way about what observations are possible, what they have time for, and what they have money for. For instance, if the available spot to observe a gesturing gorilla is four meters away, the researcher may not be able to note the animal's facial expressions because the faces of gorillas are very dark and hard to see. With a lighter-faced animal, like a chimpanzee, four meters would be no problem. In some cases, this partial gathering of information won't matter, but it's possible the facial expressions accompanying the gestures would alter the conclusions. There are other practical considerations as well. Tomasello observed that it would be interesting to study throwing in chimpanzees, but it's not something that any researcher is willing to do—if they reward throwing behavior, the chimpanzees will start throwing their feces at the researchers.

3. D. A. Leavens, W. D. Hopkins, "The Whole-Hand Point."

4. Orangutans are quite cooperative, as are bonobos, which raises the possibility that we didn't evolve to become cooperative from being noncooperative, but that chimpanzees evolved away from this trait.

5. Tomasello and Hare also ran the experiment with dogs, and the canines had no problem interpreting the cooperative pointing. The researchers attribute the dogs' sensitivity to the human-behavior agenda to their domestication.

6. E. S. Savage-Rumbaugh, "Why Are We Afraid of Apes with Languages?"

7. E. S. Savage-Rumbaugh, S. Shanker, T. J. Taylor, *Apes, Language, and the Human Mind.*

8. T. M. Pearce, "Did They Talk Their Way Out of Africa?"

9. S. Goldin-Meadow et al., "Explaining Math: Gesturing Lightens the Load."

10. S. Özçalışkan, S. Goldin-Meadow, "Gesture Is at the Cutting Edge of Early Language Development."

11. J. M. Iverson, S. Goldin-Meadow, "Gesture Paves the Way for Language Development."

12. S. Goldin-Meadow, "What Language Creation in the Manual Modality Tells Us About the Foundations of Language."

13. The children were asked to solve problems like $4+5+3=\underline{?}+3$. The adults were asked to solve problems like $x^2-5x+6=(\underline{})(\underline{})$. S. Goldin-Meadow et al., "Explaining Math: Gesturing Lightens the Load."

14. S. Goldin-Meadow, S. M. Wagner, "How Our Hands Help Us Learn."

15. In recent years, linguists have studied two very interesting cases where small deaf communities invented a sign language, the first in Nicaragua and the second among the Al-Sayyid Bedouin group in Israel. In both cases, the inception of the language has been pinpointed in time, and the codification of grammar in ensuing generations has been traced. The resulting syntactic conventions are taken as evidence of innate linguistic structure. These investigations are fascinating and important, but whether they reveal innate properties of language is considered controversial. The most salient criticism is that the deaf individuals are communicating with people who already have language. Surely the success or failure of the interpretations made by listeners who are not deaf (including, in the case of the Al-Sayyid Bedouin group, all of the deaf individuals' parents) guides the way the sign language evolves. These issues, which also relate to the investigation of homesign, are yet to be resolved. (Also see the comments of Michael Arbib and Simon Kirby in the epilogue.)

Chapter 8. You have speech
1. E. Balaban, M. A. Teillet, N. Le Douarin, "Application of the Quail-Chick Chimera System to the Study of Brain Development and Behavior."

2. S. Nadis, "Look Who's Talking."

3. According to Ramon Ferrer i Cancho (see chapter 15), the statistical analysis of higher order entropies is not statistically accurate. The kind of analysis carried out by Doyle and McCowan gives false "orders" when the data sample is not large, as is the case with their study. Ferrer i Cancho says that the conclusions drawn by Doyle and colleagues are not necessarily wrong, but more work is needed to make them really strong.

4. Incidentally, researchers have shown that humans consolidate spoken language during sleep. It's known that many different memory tasks are improved by sleeping, and the complications of speech are no exception. Scientists from the University of Chicago showed that subjects who were trained to recognize a small set of words were also able to better recognize a set of novel words that contained the same sounds as the training set. The test subjects' performance was excellent after training but declined with time. After sleep, it completely recovered. Other researchers have monitored the brain of songbirds during sleep and discovered that the parts of the brain activated during singing while awake were reactivated during sleep, suggesting that in the way we dream of speech, songbirds dream of singing.

5. W. T. Fitch, "Comparative Vocal Production and the Evolution of Speech: Reinterpreting the Descent of the Larynx."

6. T. Nishimura et al., "Descent of the Larynx in Chimpanzee Infants."

7. W. T. Fitch, "The Evolution of Speech."

8. Ibid.

9. According to Lieberman, it's been shown that when children learn American English, boys round their lips in an attempt to lengthen their vocal tracts and make their voices sound deeper. Girls pull their lips back over their teeth, making their voices higher pitched.

10. F. Ramus et al., "Language Discrimination by Human Newborns and by Cotton-Top Tamarin Monkeys."

11. R. Tincoff et al., "The Role of Speech Rhythm in Language Discrimination: Further Tests with a Non-Human Primate."

Chapter 9. You have structure

1. ". . . [T]here are, if anything, more data available to the neonate than is strictly required for phonological acquisition." P. Carr, "Scientific Realism, Sociophonetic Variation, and Innate Endowments in Phonology." April McMahon quoted Carr in a presentation about the evolution of phonology at the 2004 Evolution of Language conference in Leipzig.

2. K. Zuberbühler, "A Syntactic Rule in Forest Monkey Communication."

3. At the 2006 Rome Evolution of Language conference, Seyfarth joked that the size of an animal's vocal repertoire is best predicted by how long a scientist has been studying its species.

4. The term "syntactic nuts" originated with Peter Culicover, professor and chair of linguistics and director, Center for Cognitive Science, Ohio State University.

5. S. Pinker, R. Jackendoff, "The Faculty of Language: What's Special About It?" 15–16.

6. P. W. Culicover, R. Jackendoff, *Simpler Syntax*.

7. In addition, in the mainstream view idioms (where meaning is more than a combination of the separate meaning of the words: "She laughed her head off," "He hit it out of the park," "He had a cow") would be

considered peripheral, but Jackendoff believes that idioms and the special structural tools they offer are as important to language as basic ordering of words.

8. J. H. McWhorter, *The Power of Babel,* 188.

9. P. W. Culicover, R. Jackendoff, *Simpler Syntax,* 541.

10. E. Pennisi, "Speaking in Tongues."

11. T. W. Deacon, *The Symbolic Species,* 71. Especially important, says Deacon, is not to "underestimate what can be represented by non-symbolic means," 397.

12. R. Jackendoff, *Foundations of Language,* 253.

13. D. A. Schwartz, C. Q. Howe, D. Purves, "The Statistical Structure of Human Speech Sounds Predicts Musical Universals."

14. M. D. Hauser, J. McDermott, "The Evolution of the Music Faculty."

15. At the 2002 Harvard Evolution of Language conference, Jelle Atema, a research fellow in the Department of Cognitive and Neural Systems at Boston University, entertained attendees by playing a facsimile of a Neanderthal flute.

16. Motherese, or infant-directed speech, is characterized by lots of swings between high and low pitches, short statements, and repeated vowels. It is one of the few true universals in language; all humans do it the same way, no matter what language they speak. For this reason, motherese has been proposed as a candidate language fossil. Instead of analyzing it only as an adaptation to support a child's comprehension, the linguist Elizabeth Peters at Florida State University says it's plausible that motherese is a descendant of our ancestors' proto-language. Steven Mithen, who wrote *The Singing Neanderthals: The Origins of Music, Language, Mind and Body,* advocates a return to ideas promoted by Rousseau, Darwin, and others; specifically, that modern language was preceded by a holistic, musical protolanguage. Says Mithen, this stage of linguistic evolution helps explain phenomena like the "inherent musicality of infants." Robin Dunbar is another scholar who has written a detailed account of why language evolved with respect o music. In *Grooming, Gossip, and the Evolution of Language,* Dunbar proposes that language evolved to facilitate social bonding. Interestingly, he proposes a stage where linguistic sophistication was preceded by group chorusing.

17. In a 2003 interview Trehub spoke about how, in addition to parents, groups like the Taliban understood the power of music. "Those in charge have always known the power of music, which is why they've sought to control it one way or the other," she said. The Taliban banned music in Afghanistan, and in doing so, she explained, they removed the potential of others to stir the emotions of the population. In such emotion, a revolt could begin.

Chapter 10. You have a human brain

1. Except for the basal ganglia and a very small piece of the occipital lobe.

2. For all children who undergo hemispherectomy because of a seizure disorder, postoperative progress

depends on many factors, including whether the seizures have been brought under control. In Lacy's case, as happens sometimes for other children, a second operation was required to remove a small remaining piece of tissue that continued to cause seizures.

3. E. Bates, F. Dick, "Beyond Phrenology: Brain and Language in the Next Millennium."

4. E. Bates, "Comprehension and Production in Early Language Development."

5. S. Knecht et al., "Degree of Language Lateralization Determines Susceptibility to Unilateral Brain Lesions."

6. P. Lieberman, "On the Nature and Evolution of the Neural Bases of Human Language," 38.

7. Ibid., 38–39.

8. Ibid., 57–58.

9. Elizabeth Bates died in 2003.

10. E. Bates, "Construction Grammar and Its Implications for Child Language Research."

11. E. Bates, F. Dick, "Language, Gesture, and the Developing Brain."

12. E. Bates, F. Dick, "Beyond Phrenology: Brain and Language in the Next Millennium."

13. E. Bates, "Comprehension and Production in Early Language Development."

14. Using imaging to resolve questions about the online processing of tiny increments of language is a new and controversial field.

15. L. K. Tyler, W. D. Marslen-Wilson, E. A. Stamatakis, "Differentiating Lexical Form, Meaning, and Structure in the Neural Language System."

16. W. D. Marslen-Wilson, L. K. Tyler, "The Lexicon, Grammar, and the Past Tense."

17. The authors say, "[E]vidence for differentiation of function in the adult brain is in no way evidence per se against an emergentist view."

18. K. D. Long, G. Kennedy, E. Balaban, "Transferring an Inborn Auditory Perceptual Predisposition with Interspecies Brain Transplants."

19. Y. Kozorovitskiy et al., "Experience Induces Structural and Biochemical Changes in the Adult Primate Brain."

20. S. L. Williams, K. E. Brakke, E. S. Savage-Rumbaugh, "Comprehension Skills of Language-Competent and Nonlanguage-Competent Apes," 314.

21. E. S. Savage-Rumbaugh, R. Lewin, *Kanzi: The Ape at the Brink of the Human Mind.*

22. E. S. Savage-Rumbaugh, S. Shanker, T. J. Taylor, *Apes, Language, and the Human Mind.*

23. The relationship between the size of your brain and what you eat is a very interesting one. Said Lori Marino: "The species that are the most highly encephalized tend to be the ones that have a more complex dietary strategy, and hunting is a very complex dietary strategy, much more so than picking leaves off a tree. Even in primates, if you take two monkey species that are similar in many ways, but one is an insectivore and one is a foliavore, the insectivore will tend to be more highly encephalized. It could be that there is

something about carnivory or, more generally, complexity in dietary strategy that requires a bigger brain. Eating leaves can be complex, depending upon the kind of information that you have to process about the leaves. Eating fruits can be complex. But eating other animals is probably the most cognitively demanding, because you have to process information about the changing behavior of another animal, and you can get into arms races where the prey changes because of the predator, and then the predator changes to match that—and it ratchets up. Being a hunter doesn't mean that you're going to develop a language but as a hunter you start having more complicated things to talk about . . . and if you have human language you're probably going to be a carnivore not an herbivore."

24. T. W. Deacon, *The Symbolic Species,* 214.

25. C. Cantalupo, W. D. Hopkins, "Asymmetric Broca's Area in Great Apes."

26. E. Bates, F. Dick, "Language, Gesture, and the Developing Brain."

Chapter 11. Your genes have human mutations

1. C. S. L. Lai et al., "A Forkhead-Domain Gene Is Mutated in a Severe Speech and Language Disorder."

2. C. Knight, M. Studdert-Kennedy, J. R. Hurford "Language: A Darwinian Adaptation?" in *The Evolutionary Emergence of Language,* 5.

3. The notion of a language-specific gene echoes the idea from chapter 1 that a little speaker or linguist (a homunculus) is inside our heads and it generates and interprets language for us.

4. S. Olsen, *Mapping Human History.*

5. S. Wells, *The Journey of Man.*

6. This kind of gene is often connected with changes at the level of the whole organism.

7. W. Shu et al., "Altered Ultrasonic Vocalization in Mice with a Disruption in the Foxp2 Gene."

8. T. E. Holy, Z. Guo, "Ultrasonic Songs of Male Mice."

III. What Evolves?

1. With all due respect to Neal Stephenson, whose third book in the excellent, epic trilogy *The Baroque Cycle* is titled *The System of the World.* Stephenson's book draws its title from the third volume of Newton's *Principia Mathematica, De Mundi Systemate* (On the System of the World).

Chapter 12. Species evolve

1. In 1997, a team led by Svante Pääbo announced it had compared the mtDNA of a Neanderthal with a modern human. The two examples of DNA were so different that they suggest we are a completely different species. Keep in mind the whole genome would need to be compared for us to conclude fully that there is no Neanderthal DNA in the modern human genome.

2. A. Brumm et al., "Early Stone Technology on Flores and Its Implications for *Homo floresiensis.*"

3. Since the initial announcement of the *Homo floresiensis* discovery, there has been considerable controversy of the classification of this creature. Some scientists claim that hobbits were not our cousins in the way that Neanderthals were but that they are the ancient remains of essentially modern humans who were either pygmies (who are known to inhabit the islands) or who suffered from a disease that stunted their growth.

4. P. Mellars, "Why Did Modern Human Populations Disperse from Africa ca. 60,000 Years Ago?"

5. Ibid.

6. S. L. Salzberg et al., "Microbial Genes in the Human Genome: Lateral Transfer or Gene Loss?"

7. H. Teotónio, M. R. Rose, "Variation in the Reversibility of Evolution."

8. R. Dawkins, *The Ancestor's Tale,* 67.

9. Ibid., 75.

10. It is not clear whether average difference in DNA or expression difference is the more important. In fact, Pääbo and colleagues argued in a 2005 *Science* paper that it is a moot point—both phenomena have undoubtedly been important, and most expression differences go back to sequence differences in regulatory sequences or regulatory genes anyway.

11. M. Cáceres et al., "Elevated Gene Expression Levels Distinguish Human from Non-Human Primate Brains."

12. In DNA, the four bases, A, G, C, and T, are always paired together. T is paired with A, and G is paired with C. There are approximately 3.2 billion of these base pairs in the human genome. Of these 3.2 billion base pairs, said Enard, only 5 percent carry information that's relevant to the organism. Moreover, there is roughly 0.08 percent difference between all people, so we are all more or less equally related. (In contrast, any two individual orangutans differ much more from each other.)

13. An alternative possibility is that it resulted from the relaxation of a constraint—i.e., do other animals need it in the form they have and we don't?

14. Another reason that it is tempting to look for a single genetic mutation to explain language and culture is that although we looked the same from about 150,000 years on, the behaviors that we recognize as modern do not *appear* to have sprung forth for a long time, another 100,000 years or so. Perhaps a mutation caused radical rewiring of the brain, and therefore changes in behavior and culture, without necessarily modifying our appearance? There are many reasons why this may not be the case. We may simply have not found the evidence showing that modern culture did begin to emerge around the 150,000-year mark. Moreover, we know that a number of mutations have taken place in this time frame, although we don't currently have as much information about their effects as the effects of FOXP2. To say that any one genetic change is responsible for all of language and culture is to ignore our broad evolutionary platform.

Chapter 13. Culture evolves

1. E. S. Savage-Rumbaugh, R. Lewin, *Kanzi*.

2. K. J. Hockings, J. R. Anderson, T. Matsuzawa, "Road Crossing in Chimpanzees: A Risky Business."

3. Hunting is probably one of the most baffling examples of complicated, coordinated animal behavior that occurs without the shared planning that takes place in language. Many animals hunt, including our closest relatives, the chimpanzees. How do pack animals act independently but together in the absence of the clear imperatives humans would issue? (Imagine a group of ten humans bringing down a rearing zebra without any verbal communication between them whatsoever.) Some researchers suspect that the "order" we observe in, say, a lion attack is an emergent outcome of individuals acting alone.

4. J. Mercader, M. Panger, C. Boesch, "Excavation of a Chimpanzee Stone Tool Site in the African Rainforest."

5. J. Mercader et al., "4,300-Year-Old Chimpanzee Sites and the Origins of Percussive Stone Technology."

6. Carel van Schaik, "Why Are Some Animals So Smart?"

7. John Locke and Barry Bogin explored the importance of life stages of the individual for language evolution. They identify four developmental stages that humans pass through before adulthood, and they link different functions of language to these stages. Only humans pass through all four stages; they are infancy (birth to three years), childhood (three years to approximately six),

juvenility (sexual immaturity but independence of others for survival), and adolescence (sexual maturity). The result, say Locke and Bogin, is that the entire course of a human life in its many different phases (rather than, say, just infancy, adolescence, then adulthood) is essential for the evolution of language over time. One consequence of this is that tracking the origin of different life stages in the fossil record can reveal the course of language evolution. See J. L. Locke, B. Bogin, "Language and Life History."

8. T. W. Deacon, *The Symbolic Species*, 112.

9. Ibid., 110.

10. The opposite of a compositional utterance is a holistic one. Said Kirby: "In some sense, noncompositional language doesn't even have words, or at least not meaningful ones. Compare the following (noncompositional versus compositional):

Hi versus *I greet you*
Chutter versus *I thought I saw a pussy cat*
Went versus *walked*
Bought the farm versus *Ceased to live*"

There is a related debate within the discipline about whether language may have first been holistic and then become compositional or whether it began as separate units that could be combined. It's an interesting question for which there are no data.

11. Further research could show how complementary the two approaches are.

12. M. H. Christiansen, S. Kirby, *Language Evolution*, chapter 15.

13. Ibid., 277.

14. For more on emergent systems read Steven Johnson's *Emergence* and Kevin Kelly's *Out of Control.*

15. AIBO owners and researchers all over the world mourned in early 2006 when Sony decided to cancel production of the robot dog. The QRIO was canceled at the same time.

Chapter 14. Why things evolve

1. N. Chomsky et al., *On Nature and Language.*

2. T. W. Deacon, *The Symbolic Species,* 184.

3. Ibid., 147.

4. Ibid., 224.

5. Ibid.

6. J. Diamond, P. Bellwood, "Farmers and Their Languages: The First Expansions."

7. An odd but interesting line of inquiry into the prehistory of language is the work done on click speech sounds that are found in many African languages. Some researchers contend that clicks are extremely ancient in origin and that click languages of today all descended from one of the oldest, if not the first, human language. But there are many issues to be resolved, including the fact that much inspiration for looking at clicks in this way comes from a linguistic analysis carried out by Stanford linguist Joseph Greenberg. Greenberg grouped all click languages together into one language family, yet subsequent analyses have claimed there is little relationship between many of these languages. Greenberg's classification of the world's languages, which Luigi Cavalli-Sforza also used in his attempt to trace the history of languages and genes recounted in *Genes, Peoples, and Languages,* is regarded as highly controversial, if not flat-out wrong, by many linguists today.

IV. Where Next?

Chapter 15. The future of the debate

1. D. Bickerton, "Language Evolution: A Brief Guide for Linguists."

2. Researchers such as Irene Pepperberg, Sue Savage-Rumbaugh, and Phil Lieberman had long proposed that behaviors in other animals, whether they are learned or whether they arise naturally, expand our understanding of language evolution. Moreover, Chomskyan linguistics, if not Chomsky himself, with its emphasis on the innate and uniquely human language capacity, had for many years discouraged researchers from looking to animals for information about human language.

3. Instigated by the Hauser, Chomsky, Fitch *Science* paper, but not necessarily about the specific hypotheses they proposed.

4. G. Origgi, D. Sperber, "A Pragmatic Perspective on the Evolution of Language and Languages."

5. In a presentation at the 2005 Evolution of Language Symposium at Stony Brook, Sperber talked about the human ability to construct representations of representations— that is, metarepresentations. "A metarepresentational ability," he said, "need not have communication as a

primary function, but it makes inferential communication possible." He elaborated on this by describing the way the briefest moment of nonlinguistic communication can be built from a complicated layering of inference and shared understanding. For example, when Mary and her son Peter take a walk through the forest they pass a shrub with a kind of berry unknown to Peter, Mary establishes eye contact with him, bites a berry and spits it out, and from this Peter understands that she means that he should not eat these berries. Such communication involves the following layers of representation and intention: "1. Mary picks a berry, knowing that they are inedible, and spits it out; 2. Peter looks at her and infers from what she is doing that she thinks that these berries are inedible, and concludes that they are inedible; 3. Mary intended that Peter would draw this conclusion (it was her intention to inform Peter); 4. Peter infers from the fact that Mary had established eye contact with him that it was her intention that he would draw this conclusion; 5. Mary intended that Peter would infer from her having established eye contact with him that it was her intention that he should draw this conclusion (it was her intention not just to inform Peter but to communicate with him—that is, to establish mutual understanding)."

6. Chris Knight proposes that a revolutionary cultural shift was marked by the use of ocher for body adornment between 130,000 and 70,000 years before the present. He argues that ocher, the color of blood,

was used by coalitions of females to disguise evidence of fertility. From these symbolic beginnings, culture emerged. Knight talks about the human revolution, but he doesn't define it as a sudden event that explains all of language. Rather, he says, the defining feature of a revolution is that it turns the world upside down.

7. N. Chomsky et al., *On Nature and Language.*

8. Q. Wen, D. B. Chklovskii, "Segregation of the Brain into Gray and White Matter."

9. A. Marantz, Y. Miyashita, W. O'Neil, *Image, Language, Brain,* 23.

10. http://www.derekbickerton.com/.

11. In the words of Pinker and Jackendoff, after E. O. Wilson.

12. C. Yang, "Dig-Dig, Think-Thunk."

13. E. S. Savage-Rumbaugh, R. Lewin, *Kanzi,* chapter 7.

Chapter 16. The future of language and evolution

1. D.-E. Nilsson, S. Pelger, "A Pessimistic Estimate of the Time Required for an Eye to Evolve."

2. Dan Dediu, Simon Kirby's Ph.D. student, and the phonologist Bob Ladd have found a significant correlation between these two genes and the presence or absence of tone in a language, such as Chinese. They argue that the recent variant of the genes, which most Europeans possess, makes tone languages less likely. It's possible that possessing one of these genes means that learning or producing a tone is more difficult. Said Kirby, "If correct, this will be the first time that a

genetic difference has been shown to make a difference in the language faculty such that it changes the structure of the world's languages."

3. H. Stefansson et al., "A Common Inversion Under Selection in Europeans."

4. Nicholas Wade, *New York Times,* Tuesday, March 7, 2006.

5. Does this mean that language is a mechanism of evolution in the same way that sexual and asexual reproduction are—that is, a device that changes the status of evolutionary process? If this were true, it would mean that there is something very important about language (and saying this would not be simply a case of anthropocentrism in the same way that, to continue an analogy by Steven Pinker and Richard Dawkins, if we were all elephants, this book would be exploring "trunkitude" as the accomplishment of evolution. Trunks are unique features but they are not evolutionary mechanisms).

6. F. Dyson, "Make Me a Hipparoo."

7. In "Language Evolution: A Brief Guide for Linguists," Derek Bickerton argues that language and human evolution have stopped. He writes, "Of course it [language evolution] has stopped, because the biological development of humans (saving the odd minor development like the spread of lactose tolerance or proneness to sickle-cell anemia) has, to all intents and purposes, stopped also. What is happening (and has been happening for perhaps as many as a hundred thousand years) is cultural change (sometimes misleadingly described as 'cultural evolution'); within the envelope of the language faculty, languages are recycling the limited alternatives that this biological envelope makes available . . . language evolution and changes in languages operate on different time-scales, involve different factors, and follow different courses to different ends (or rather, to the end of a complete language faculty in the first case and to no particular end in the second). To muddle them merely confuses an already sufficiently confused field."

8. W. J. Sutherland, "Parallel Extinction Risk and Global Distribution of Languages and Species."

9. See T. W. Deacon, *The Symbolic Species,* for a discussion of this topic.

10. Ibid., chapter 14.

11. Ibid.

BIBLIOGRAPHY

Aitchison, Jean, *The Seeds of Speech: Language Origin and Evolution* (Cambridge, England: Canto, 2000).

Alford, Bobby R., "Core Curriculum Syllabus: Review of Anatomy—the Larynx," http://www.bcm.edu/oto/studs/anat/larynx.htm, accessed 2006.

Arbib, Michael A., "From Monkey-Like Action Recognition to Human Language: An Evolutionary Framework for Neurolinguistics," *Behavioral and Brain Sciences* 28 (2005): 105–24.

Arnold, Kate, and Klaus Zuberbühler, "Semantic Combinations in Primate Calls," *Nature* 441 (2006): 303.

Baker, Mark C., *The Atoms of Language* (New York: Basic Books, 2001).

Balaban, E., M. A. Teillet, and N. Le Douarin, "Application of the Quail-Chick Chimera System to the Study of Brain Development and Behavior," *Science* 241 (1988): 1339–42.

Balter, Michael, "Are Human Brains Still Evolving? Brain Genes Show Signs of Selection," *Science* 309 (2005): 1662–63.

Baronchelli, Andrea, Maddalena Felici, Vittorio Loreto, Emanuele Caglioti, and Luc Steels, "Sharp Transition Towards Shared Vocabularies in Multi-Agent Systems," *Journal of Statistical Mechanics: Theory and Experiment* 2006 (2006): P06014.

Bates, Elizabeth, "Comprehension and Production in Early Language Development: Comments on Savage-Rumbaugh et al.," *Monographs of the Society for Research in Child Development* 58 (1993): 222–42.

———, "Construction Grammar and Its Implications for Child Language Research," *Journal of Child Language* 25 (1998): 443–84.

———, "Plasticity, Localization and Language Development," in Sarah H. Broman and Jack Fletcher, eds., *The Changing Nervous System: Neurobehavioral Consequences of Early Brain Disorders* (New York: Oxford University Press, 1999), pp. 214–53.

Bates, Elizabeth, and Frederic Dick, "Beyond Phrenology: Brain and Language in the Next Millennium," *Brain and Language* 71 (2000): 18–21.

———, "Language, Gesture, and the Developing Brain," *Developmental Psychobiology* 40 (2002): 293–310.

Bates, E., F. Dick, and B. Wulfeck, "Not So Fast: Domain-General Factors Can Account for Selective Deficits in Grammatical Processing," *Behavioral and Brain Sciences* 22 (1999): 96–97.

Bauer, Laurie, and Peter Trudgill, *Language Myths* (New York: Penguin Books, 1998).

Bekoff, Marc, Colin Allen, and Gordon M. Burghardt, *The Cognitive Animal: Empirical and Theoretical Perspectives on Animal Cognition* (Cambridge, Mass.: MIT Press, 2002).

Belpaeme, Tony, Bart de Boer, and Paul Vogt, "Modelling Language Origins and Evolution," paper presented at Fifth Evolution of Language Conference, Leipzig, Germany, March 2004.

Ben-Ari, Elia T. "A Throbbing in the Air," *BioScience* 49 (1999): 353–58.

Beran, M. J., D. M. Rumbaugh, and E. S. Savage-Rumbaugh, "Chimpanzee (Pan Troglodytes) Counting in a Computerized Testing Paradigm," *The Psychological Record* 48 (1998): 3–19.

Bever, Thomas, and Mario Montalbetti, "Noam's Ark," *Science* 298 (2002): 1565–66.

Bhattacharjee, Yudhijit, "From Heofonum to Heavens," *Science* 303 (2004): 1326–28.

Bickerton, Derek, *Language & Species* (Chicago: University of Chicago Press, 1990).

————, *Language and Human Behavior* (Seattle: University of Washington Press, 1995).

————, "Chomsky: Between a Stony Brook and a Hard Place," www.derekbickerton .com, accessed 2005.

————, "Language Evolution: A Brief Guide for Linguists," *Lingua* 117 (2007): 510–26.

Blackburn, Robin, and Oliver Kamm, "For and Against Chomsky," *Prospect Magazine,* November 2005.

Blackburn, Simon, "Meet the Flintstones," *The New Republic* (November 25, 2002).

Bond, A. B., A. C. Kamil, and R. P. Balda, "Social Complexity and Transitive Inference in Corvids," *Animal Behaviour* 65 (2003): 479–87.

Brakke, K. E., and E. S. Savage-Rumbaugh, "The Development of Language Skills in Bonobo and Chimpanzee—I. Comprehension," *Language and Communication* 15 (1995): 121–48.

Brannon, Elizabeth M., and Herbert S. Terrace, "Representation of the Numerosities 1-9 by Rhesus Macaques (*Macaca mulatta*)," *Journal of Experimental Psychology: Animal Behavior Processes* 26 (2000): 31–49.

Breuer, Thomas, Mireille Ndoundou-Hockemba, and Vicki Fishlock, "First Observation of Tool Use in Wild Gorillas," *PLoS Biology* 3 (2005): e380.

Bromberger, Sylvain, "Chomsky's Revolution," *The New York Review of Books* 49 (April 25, 2002).

Brumm, Adam, Fachroel Aziz, Gert D. van den Bergh, Michael J. Morwood, Mark W. Moore, Iwan Kurniawan, Douglas R. Hobbs, and Richard Fullagar, "Early Stone Technology on Flores and Its Implications for *Homofloresiensis,*" *Nature* 441 (2006): 624–28.

Bustamante, Carlos D., Adi Fledel-Alon, Scott Williamson, Rasmus Nielsen, Melissa Todd Hubisz, Stephen Glanowski, David M. Tanenbaum, Thomas J. White, John J. Sninsky, Ryan D. Hernandez, Daniel Civello, Mark D. Adams, Michele Cargill, and Andrew G. Clark, "Natural Selection on Protein-Coding Genes in the Human Genome," *Nature* 437 (2005): 1153–57.

Cáceres, Mario, Joel Lachuer, Matthew A. Zapala, John C. Redmond, Lili Kudo, Daniel H. Geschwind, David J. Lockhart, Todd M. Preuss, and Carrolee Barlow, "Elevated Gene Expression Levels Distinguish Human from Non-Human Primate Brains," *PNAS* 100 (2003): 13030–35.

————, "Zipf's Law from a Communicative Phase Transition," *European Physical Journal B* 47 (2005): 449–57.

————, "When Language Breaks into Pieces: A Conflict Between Communication Through Isolated Signals and Language," *Biosystems* 84 (2006): 242–53.

Cangelosi, Angelo, and Domenico Parisi, *Simulating the Evolution of Language* (New York: Springer, 2002).

Cangelosi, Angelo, Andrew D. M. Smith, and Kenny Smith, eds., *The Evolution of Language: Proceedings of the Sixth International Conference,* paper (Singapore: World Scientific, 2006).

Cantalupo, Claudio, and William D. Hopkins, "Asymmetric Broca's Area in Great Apes," *Nature* 414 (2001): 505.

Carr, Philip, "Scientific Realism, Sociophonetic Variation, and Innate Endowments in Phonology," in eds. Noel Burton-Roberts, Philip Carr, and Gerard Docherty, *Phonological Knowledge* (New York: Oxford University Press, 2000), pp. 67–104.

Carruthers, Peter, Stephen Laurence, and Stephen P. Stich, eds., *The Innate Mind: Structure and Contents* (New York: Oxford University Press, 2005).

Carstairs-McCarthy, Andrew, "Many Perspectives, No Consensus," *Science* 303 (2004): 1299–300.

Cattuto, Ciro, Vittorio Loreto, and Luciano Pietronero, "Semiotic Dynamics and Collaborative Tagging," *PNAS* 104 (2007): 1461–64.

Cavalli-Sforza, L. L., *Genes, Peoples, and Languages* (Berkeley: University of California Press, 2001).

Cheney, Dorothy L., and Robert M. Seyfarth, *Baboon Metaphysics: The Evolution of a Social Mind* (Chicago: University of Chicago Press, 2007).

————, *How Monkeys See the World: Inside the Mind of Another Species* (Chicago: University of Chicago Press, 1990).

Chomsky, Noam, "A Review of B. F. Skinner's Verbal Behavior," *Language* 25 (1959): 26–58.

————, "A Review of B. F. Skinner's Verbal Behavior," in eds. Leon A. Jakobovits and Murray S. Miron, *Readings in the Psychology of Language* (Englewood, N.J.: Prentice-Hall, 1967), pp. 142–43.

————, *Problems of Knowledge and Freedom* (New York: Pantheon Books, 1971).

————, "The Case against B. F. Skinner," *The New York Review of Books* 17 (December 30, 1971).

————, *Language and Mind* (New York: Harcourt Brace Jovanovich, 1972).

————, "Chomsky Replies," *The New York Review of Books* 20 (July 19, 1973).

————, *Reflections on Language* (New York: Pantheon Books, 1975).

————, *Rules and Representations* (London: B. Blackwell, 1980).

————, "Discussion of Putnam's Comments," n ed. M. Piattelli Palmarini, *Language and Learning: The Debate Between Jean Piaget and Noam Chomsky* (Cambridge, Mass.: Harvard University Press, 1982), p. 321.

————, *Language and Problems of Knowledge: The Managua Lectures* (Cambridge, Mass.: MIT Press, 1988).

————, "Universal Grammar," *The New York Review of Books* 38 (December 19, 1991).

————, "Language and Evolution," *The New York Review of Books* 43 (February 1, 1996).

————, *New Horizons in the Study of Language and Mind* (New York: Cambridge University Press, 2000).

————, "Chomsky's Revolution: An Exchange," *The New York Review of Books* 49 (July 18, 2002).

Chomsky, Noam, Adriana Belletti, and Luigi Rizzi, *On Nature and Language* (New York: Cambridge University Press, 2002).

Chomsky, Noam, Marc D. Hauser, and W. Tecumseh Fitch, "Appendix. The Minimalist Program," http://www.wjh.harvard.edu/~mnkylab/publications/recent.htm.

Chomsky, Noam, and Mitsou Ronat, *Language and Responsibility* (Hassocks, England: Harvester Press, 1979).

Christiansen, Morten H., and Christopher M. Conway, "The Importance of Hierarchical Learning: A Computational Study of Sequential Learning in Human and Non-Human Primates," paper presented at Evolution of Language conference, Harvard University, Cambridge, Mass., March 2002.

Christiansen, Morten H., and Simon Kirby, *Language Evolution,* Studies in the Evolution of Language, vol. 3 (New York: Oxford University Press, 2003).

————, "Language Evolution: Consensus and Controversies," *Trends in Cognitive Sciences* 7 (2003): 300–307.

Clark, G. A., "Neandertal Archaeology: Implications for Our Origins," *American Anthropologist* 104 (2002): 50–67.

Clarke, Esther, Ulrich H. Reichard, and Klaus Zuberbühler, "The Syntax and Meaning of Wild Gibbon Songs," *PLoS One* (2006): e73.

Coppola, Marie, and Elissa L. Newport, "Grammatical Subjects in Home Sign: Abstract Linguistic Structure in Adult Primary Gesture Systems Without Linguistic Input," *PNAS* 102 (2005): 19249–53.

Corballis, Michael C., *From Hand to Mouth: The Origins of Language* (Princeton: Princeton University Press, 2002).

————, "How Gestures Became Vocal," paper presented at Evolution of Language conference, Harvard University, Cambridge, Mass., March 2002.

————, "From Mouth to Hand: Gesture, Speech, and the Evolution of Right-Handedness," *Behaviorial and Brain Sciences* 26 (2003): 199–260.

————, "Did Language Evolve Before Speech?" 2005, http://www.linguistics.stonybrook.edu/events/nyct05/index_files/Page408.htm.

Cruse, D. A., *Lexical Semantics*, Cambridge Textbooks in Linguistics (New York: Cambridge University Press, 1986).

Culicover, Peter W., and Ray Jackendoff, *Simpler Syntax* (New York: Oxford University Press, 2005).

Culotta, Elizabeth, and Brooks Hanson, "First Words," *Science* 303 (2004): 1315.

Darwin, Charles, *On the Origin of Species* (Cambridge: Harvard University Press, 1964).

————, *The Descent of Man* (Amherst, N.Y.: Prometheus Books, 1998).

Dave, Amish S., and Daniel Margoliash, "Song Replay During Sleep and Computational Rules for Sensorimotor Vocal Learning," *Science* 290 (2000): 812–16.

Dawkins, Richard, *The Ancestor's Tale: A Pilgrimage to the Dawn of Evolution* (Boston: Houghton Mifflin, 2004).

de Waal, F. B. M., *Chimpanzee Politics: Power and Sex among Apes* (Baltimore: Johns Hopkins University Press, 1998).

———, *The Ape and the Sushi Master: Cultural Reflections by a Primatologist* (New York: Basic Books, 2001).

de Waal, F. B. M., and Peter L. Tyack, *Animal Social Complexity: Intelligence, Culture, and Individualized Societies* (Cambridge, Mass.: Harvard University Press, 2003).

———, "Cultural Primatology Comes of Age," *Nature* 399 (1999): 635–36.

———, "Do Humans Alone 'Feel Your Pain'?" *The Chronicle of Higher Education* (October 26, 2001).

Deacon, Terrence William, *The Symbolic Species: The Co-Evolution of Language and the Brain* (New York: W.W. Norton, 1997).

Dehaene, Stanislas, Veronique Izard, Pierre Pica, and Elizabeth Spelke, "Core Knowledge of Geometry in an Amazonian Indigene Group," *Science* 311 (2006): 381–84.

Dehaene, Stanislas, Lionel Naccache, Laurent Cohen, Denis Le Bihan, Jean-François Mangin, Jean-Baptiste Poline, and Denis Rivière, "Cerebral Mechanisms of Word Masking and Unconscious Repetition Priming," *Nature Neuroscience* 4 (2001): 752–58.

Dennett, Daniel Clement, *Darwin's Dangerous Idea: Evolution and the Meanings of Life* (New York: Simon & Schuster, 1996).

———, " 'Darwinian Fundamentalism': An Exchange," *The New York Review of Books* 44 (August 14, 1997).

Dennett, Daniel C., and John Maynard Smith, " 'Confusion over Evolution': An Exchange," *The New York Review of Books* 40 (January 14, 1993).

Diamond, Jared, and Peter Bellwood, "Farmers and Their Languages: The First Expansions," *Science* 300 (2003): 597–603.

Dick, F., and E. Bates, "Grodzinsky's Latest Stand—or, Just How Specific Are 'Lesion-Specific' Deficits?" *Behavioral and Brain Sciences* 23 (2000): 29.

Dunbar, R. I. M., *Grooming, Gossip, and the Evolution of Language* (Cambridge, Mass.: Harvard University Press, 1996).

Dunn, Michael, Angela Terrill, Ger Reesink, Robert A. Foley, and Stephen C. Levinson, "Structural Phylogenetics and the Reconstruction of Ancient Language History," *Science* 309 (2005): 2072–75.

Dyson, Freeman, "Make Me a Hipparoo," *New Scientist* (February 11, 2006).

Ehret, Gunter, and Sabine Riecke, "Mice and Humans Perceive Multiharmonic Communication Sounds in the Same Way," *PNAS* 99 (2002): 479–82.

Elango, Navin, James W. Thomas, and Soojin V. Yi, "Variable Molecular Clocks in Hominoids," *PNAS* 103 (2006): 1370–75.

Enard, Wolfgang, Philipp Khaitovich, Joachim Klose, Sebastian Zollner, Florian Heissig, Patrick Giavalisco, Kay Nieselt-Struwe, Elaine Muchmore, Ajit Varki, Rivka Ravid, Gaby M. Doxiadis, Ronald E. Bontrop, and Svante Pääbo, "Intra- and Interspecific Variation in Primate Gene Expression Patterns," *Science* 296 (2002): 340–43.

Enard, Wolfgang, Molly Przeworski, Simon E. Fisher, Cecilia S. L. Lai, Victor Wiebe, Takashi Kitano, Anthony P. Monaco, and Svante Pääbo, "Molecular Evolution of Foxp2, a Gene Involved in Speech and Language," *Nature* 418 (2002): 869–72.

Evans, Patrick D., Sandra L. Gilbert, Nitzan Mekel-Bobrov, Eric J. Vallender, Jeffrey R. Anderson, Leila M. Vaez-Azizi, Sarah A. Tishkoff, Richard R. Hudson, and Bruce T.

Lahn, "*Microcephalin,* a Gene Regulating Brain Size, Continues to Evolve Adaptively in Humans," *Science* 309 (2005): 1717–20.

Fenn, Kimberly M., Howard C. Nusbaum, and Daniel Margoliash, "Consolidation During Sleep of Perceptual Learning of Spoken Language," *Nature* 425 (2003): 614–16.

Ferrer i Cancho, Ramon, "The Variation of Zipf's Law in Human Language," *European Physical Journal B* 44 (2005): 249–57.

Ferrer i Cancho, Ramon, Oliver Riordan, and Béla Bollobás, "The Consequences of Zipf's Law for Syntax and Symbolic Reference," *Proceedings of The Royal Society of London. Series B, Biological Sciences* online (2005): 1–5.

Ferrer i Cancho, Ramon, and Ricard V. Sole, "Two Regimes in the Frequency of Words and the Origins of Complex Lexicons: Zipf's Law Revisited," *Journal of Quantitative Linguistics* 8 (2001): 165–73.

———, "Least Effort and the Origins of Scaling in Human Language," *PNAS* 100 (2003): 788–91.

Fisher, Simon E., and Gary F. Marcus, "The Eloquent Ape: Genes, Brains and the Evolution of Language," *Nature Reviews Genetics* 7 (2006): 9–20.

Fitch, W. Tecumseh, "The Biology and Evolution of Music: A Comparative Perspective," *Cognition* 100 (2006): 173–215.

———, "The Evolution of Speech: A Comparative Review," *Trends in Cognitive Sciences* 4 (2000): 258–67.

———, "Comparative Vocal Production and the Evolution of Speech: Reinterpreting the Descent of the Larynx," in ed. Alison Wray, *The Transition to Language* (Oxford: Oxford University Press, 2002), pp. xi, 410.

Fitch, W. Tecumseh, and Marc D. Hauser, "Computational Constraints on Syntactic Processing in a Nonhuman Primate," *Science* 303 (2004): 377–80.

Fitch, W. Tecumseh, Marc D. Hauser, and Noam Chomsky, "The Evolution of the Language Faculty: Clarifications and Implications," *Cognition* 97 (2005): 179–210.

Forster, Peter, and Alfred Toth, "Toward a Phylogenetic Chronology of Ancient Gaulish, Celtic, and Indo-European," *PNAS* 100 (2003): 9079–84.

Fouts, Roger, and Stephen Tukel Mills, *Next of Kin: What Chimpanzees Have Taught Me About Who We Are* (New York: William Morrow, 1997).

Friederici, Angela D., Jörg Bahlmann, Stefan Heim, Ricarda I. Schubotz, and Alfred Anwander, "The Brain Differentiates Human and Non-Human Grammars: Functional Localization and Structural Connectivity," *PNAS* 103 (2006): 2458–63.

Gardner, Martin, "Monkey Business," *The New York Review of Books* 27, 3–6 (March 20, 1980).

Gentner, Timothy Q., Kimberly M. Fenn, Daniel Margoliash, and Howard C. Nusbaum, "Recursive Syntactic Pattern Learning by Songbirds," *Nature* 440 (2006): 1204–7.

Gilad, Yoav, Alicia Oshlack, Gordon K. Smyth, Terence P. Speed, and Kevin P. White, "Expression Profiling in Primates Reveals a Rapid Evolution of Human Transcription Factors," *Nature* 440 (2006): 242–45.

Gilbert, Aubrey L., Terry Regier, Paul Kay, and Richard B. Ivry, "Whorf Hypothesis Is Supported in the Right Visual Field but Not the Left," *PNAS* 103 (2006): 489–94.

Givón, T., "The Visual Information-Processing System as an Evolutionary Precursor of Human Language," *Typological Studies in Language* 53 (2002): 3–50.

Goldin-Meadow, Susan, "Watching Language Grow," *PNAS* 102 (2005): 2271–72.

———, "What Language Creation in the Manual Modality Tells Us About the Foundations of Language," *The Linguistic Review* 22 (2005): 199–225.

Goldin-Meadow, S., H. Nusbaum, S. D. Kelly, and S. Wagner, "Explaining Math: Gesturing Lightens the Load," *Psychological Science* 12 (2001): 516–22.

Goldin-Meadow, S., and S. M. Wagner, "How Our Hands Help Us Learn," *Trends in Cognitive Science* 9 (2006): 234–41.

Gordon, Peter, "Numerical Cognition Without Words: Evidence from Amazonia," *Science* 306 (2004): 496–99.

Gould, Stephen Jay, *Full House: The Spread of Excellence from Plato to Darwin* (New York: Harmony Books, 1996).

———, "Why Darwin?" *The New York Review of Books* 43 (April 4, 1996).

———, "Darwinian Fundamentalism," *The New York Review of Books* 44 (June 12, 1997).

———, "Evolution: The Pleasures of Pluralism," *The New York Review of Books* 44 (June 26, 1997).

Graddol, David, "The Future of Language," *Science* 303 (2004): 1329–31.

Grandin, Temple, and Catherine Johnson, *Animals in Translation: Using the Mysteries of Autism to Decode Animal Behavior* (New York: Scribner, 2005).

Gray, Russell, "Pushing the Time Barrier in the Quest for Language Roots," *Science* 309 (2005): 2007–8.

Gray, Russell D., and Quentin D. Atkinson, "Language-Tree Divergence Times Support the Anatolian Theory of Indo-European Origin," *Nature* 426 (2003): 435–39.

Hacking, Ian, "Chomsky and His Critics," *The New York Review of Books* 27 (October 23, 1980).

Halder, G., P. Callaerts, and W. J. Gehring, "Induction of Ectopic Eyes by Targeted Expression of the Eyeless Gene in Drosophila," *Science* 267 (1995): 1788–92.

Hare, B., J. Call, and M. Tomasello, "Communication of Food Location Between Human and Dog (*Canis familiaris*)," *Evolution of Communication* 3 (1998): 137–59.

Hare, B., and M. Tomasello, "Chimpanzees Are More Skillful in Competitive Than in Cooperative Cognitive Tasks," *Animal Behaviour* 68 (2004): 571–81.

Harman, Gilbert, *On Noam Chomsky; Critical Essays*, Modern Studies in Philosophy (Garden City, N.Y.: Anchor Press, 1974).

Harris, Randy Allen, *The Linguistics Wars* (New York: Oxford University Press, 1993).

Harris, Roy, and Talbot J. Taylor, *Landmarks in Linguistic Thought: The Western Tradition from Socrates to Saussure* (New York: Routledge, 1989).

Hart, B. L., L. A. Hart, M. McCoy, and C. R. Sarath, "Cognitive Behaviour in Asian Elephants: Use and Modification of Branches for Fly Switching," *Animal Behaviour* 62 (2001): 839–47.

Hauser, Marc D., *The Evolution of Communication* (Cambridge, Mass.: MIT Press, 1996).

———, *Wild Minds: What Animals Really Think* (New York: Henry Holt, 2000).

Hauser, Marc D., Noam Chomsky, and W. Tecumseh Fitch, "The Faculty of Language: What Is It, Who Has It, and How Did It Evolve?" *Science* 298 (2002): 1569–79.

Hauser, Marc D, and Josh McDermott, "The Evolution of the Music Faculty: A Comparative Perspective," *Nature Neuroscience* 6 (2003): 663–8.

Hewes, Gordon W., "Primate Communication and the Gestural Origin of Language," *Current Anthropology* 14 (1973): 5–24.

Hockings, K. J., J. R. Anderson, and T. Matsuzawa, "Road Crossing in Chimpanzees: A Risky Business," *Current Biology* 16 (2006): R668–R70.

Holden, Constance, "The Origin of Speech," *Science* 303 (2004): 1316–19.

———, "Hunter-Gatherers Grasp Geometry," *Science* 311 (2006): 317.

Holy, T. E., and Z. Guo, "Ultrasonic Songs of Male Mice," *PLoS Biology* 3 (2005): 2177–86.

Hunt, G. R., and R. D. Gray, "Diversification and Cumulative Evolution in New Caledonian Crow Tool Manufacture," *Proceedings of the Royal Society of London B* 270 (2003): 867–74.

Hurford, James R., Michael Studdert-Kennedy, and Chris Knight, eds., *Approaches to the Evolution of Language: Social and Cognitive Bases* (New York: Cambridge University Press, 1998).

International Human Genome Sequencing Consortium, "Finishing the Euchromatic Sequence of the Human Genome," *Nature* 431 (2004): 931–45.

Iverson, Jana M., and Susan Goldin-Meadow, "Gesture Paves the Way for Language Development," *Psychological Science* 16 (2005): 367–71.

Jackendoff, Ray, *Foundations of Language: Brain, Meaning, Grammar, Evolution* (New York: Oxford University Press, 2002).

Jackendoff, Ray, and Fred Lerdahl, "The Capacity for Music: What Is It, and What's Special About It?" *Cognition* 100 (2006): 33–72.

Jackendoff, Ray, and Steven Pinker, "The Nature of the Language Faculty and Its Implications for Evolution of Language," *Cognition* 97 (2005): 211–25.

Janata, Petr, and Scott T. Grafton, "Swinging in the Brain: Shared Neural Substrates for Behaviors Related to Sequencing and Music," *Nature Neuroscience* 6 (2003): 682–87.

Janik, V. M., L. S. Sayigh, and R. S. Wells, "Signature Whistle Shape Conveys Identity Information to Bottlenose Dolphins," *PNAS* 103 (2006): 8293–97.

Jarvis, Erich D., Onur Güntükün, Laura Bruce, András Csillag, Harvey Karten, Wayne Kuenzel, Loreta Medina, George Paxinos, David J. Perkel, Toru Shimizu, Georg Striedter, J. Martin Wild, Gregory F. Ball, Jennifer Dugas-Ford, Sarah E. Durand, Gerald E. Hough, Scott Husband, Lubica Kubikova, Diane W. Lee, Claudio V. Mello, Alice Powers, Connie Siang, Tom V. Smulders, Kazuhiro Wada, Stephanie A. White, Keiko Yamamoto, Jing Yu, Anton Reiner, and Ann B. Butler, "Avian Brains and a New Understanding of Vertebrate Brain Evolution," *Nature Reviews Neuroscience* 6 (2005): 151–59.

Johnson, George, *Fire in the Mind: Science, Faith, and the Search for Order* (New York: Alfred A. Knopf, 1995).

Johnson, Steven, *Emergence: The Connected Lives of Ants, Brains, Cities, and Software* (New York: Scribner, 2001).

Jordan, Kerry E., and Elizabeth M. Brannon, "The Multisensory Representation of Number in Infancy," *PNAS* 103 (2006): 3486–89.

Kaessmann, H., and S. Pääbo, "The Genetical History of Humans and the Great Apes," *Journal of Internal Medicine* 251 (2002): 1–18.

Kaessmann, Henrik, Victor Wiebe, Gunter Weiss, and Svante Pääbo, "Great Ape DNA Sequences Reveal a Reduced Diversity and an Expansion in Humans," *Nature Genetics* 27 (2001): 155–56.

Kalant, Harold, Steven Pinker, and Werner Kalow, "Evolutionary Psychology: An Exchange," *The New York Review of Books* 44 (October 9, 1997).

Kaminski, Juliane, Josep Call, and Julia Fischer, "Word Learning in a Domestic Dog: Evidence for 'Fast Mapping,' " *Science* 304 (2004): 1682–83.

Kaplan, Frédéric, "Language as It Could Be: How Robots Can Create New Linguistic Phenomena," paper presented at Evolution of Language conference, Harvard University, Cambridge, Mass., March 2002.

Kasher, Asa, *The Chomskyan Turn* (Oxford: Blackwell, 1991).

Kelly, Kevin, *Out of Control: The Rise of Neo-Biological Civilization* (Reading, Mass.: Addison-Wesley, 1994).

Kenneally, Christine, "Prosody, Animacy, and Syntax in the Perception of Speech," Ph.D. dissertation, University of Cambridge, 1997.

Kenward, Ben, Alex A. S. Weir, Christian Rutz, and Alex Kacelnik, "Tool Manufacture by Naive Juvenile Crows," *Nature* 433 (2005): 121.

Khaitovich, Philipp, Ines Hellmann, Wolfgang Enard, Katja Nowick, Marcus Leinweber, Henriette Franz, Gunter Weiss, Michael Lachmann, and Svante Pääbo, "Parallel Patterns of Evolution in the Genomes and Transcriptomes of Humans and Chimpanzees," *Science* 309 (2005): 1850–54.

King, Barbara J., *The Dynamic Dance: Nonvocal Communication in African Great Apes* (Cambridge, Mass.: Harvard University Press, 2004).

King, J., D. Rumbaugh, and S. Savage-Rumbaugh, "Perception as Personality Traits and Semantic Learning in Evolving Hominids," in eds. Michael C. Corballis and S. E. G. Lea, *The Descent of Mind: Psychological Perspectives on Hominid Evolution* (New York: Oxford University Press, 1999), pp. 98–115.

Kirby, Simon, "Spontaneous Evolution of Linguistic Structure: An Iterated Learning Model of the Emergence of Regularity and Irregularity," *IEEE Transactions on Evolutionary Computations* 5 (2001): 102–10.

———, "Natural Language from Artificial Life," *Artificial Life* 8 (2002): 185–216.

———, "The Mechanisms of Adaptive Linguistic Evolution," paper presented at Morris Symposium on Language Evolution, Stony Brook, N.Y., October 2005.

Kirby, Simon, and James R. Hurford, "The Emergence of Linguistic Structure: An Overview of the Iterated Learning Model," in eds. Angelo Cangelos and Domenico Parisi, *Simulating the Evolution of Language* (London: Springer Verlag, 2002).

Klein, Richard G., "Whither the Neanderthals?" *Science* 299 (2003): 1525–27.

Knecht, S., A. Floel, B. Drager, C. Breitenstein, J. Sommer, H. Henningsen, E. B. Ringelstein, and A. Pascual-Leone, "Degree of Language Lateralization Determines Susceptibility to Unilateral Brain Lesions," *Nature Neuroscience* 5 (2002): 695–99.

Knight, Chris, "Decoding Chomsky," *European Review* 12 (2004): 581–604.

Knight, Chris, Michael Studdert-Kennedy, and James R. Hurford, eds., *The Evolutionary Emergence of Language: Social Function and the Origins of Linguistic Form* (New York: Cambridge University Press, 2000).

Kozorovitskiy, Yevgenia, Charles G. Gross, Catherine Kopil, Lisa Battaglia, Meghan McBreen, Alexis M. Stranahan, and Elizabeth Gould, "Experience Induces Structural and Biochemical Changes in the Adult Primate Brain," *PNAS* 102 (2005): 17478–82.

Krutzen, Michael, Janet Mann, Michael R. Heithaus, Richard C. Connor, Lars Bejder, and William B. Sherwin, "Cultural Transmission of Tool Use in Bottlenose Dolphins," *PNAS* 102 (2005): 8939–43.

Lai, Cecilia S. L., Simon E. Fisher, Jane A. Hurst, Faraneh Vargha-Khadem, and Anthony P. Monaco, "A Forkhead-Domain Gene Is Mutated in a Severe Speech and Language Disorder," *Nature* 413 (2001): 519–23.

Lakoff, George, "Deep Language," *The New York Review of Books* 20 (February 8, 1973).

Leavens, D. A., and W. D. Hopkins, "The Whole-Hand Point: The Structure and Function of Pointing from a Comparative Perspective," *Journal of Comparative Psychology* 113 (1999): 417–25.

Lieberman, Philip, *The Biology and Evolution of Language* (Cambridge, Mass.: Harvard University Press, 1984).

———, *Eve Spoke: Human Language and Human Evolution* (New York: W.W. Norton, 1998).

———, *Human Language and Our Reptilian Brain: The Subcortical Bases of Speech, Syntax, and Thought*, Perspectives in Cognitive Neuroscience (Cambridge, Mass.: Harvard University Press, 2000).

———, "On the Nature and Evolution of the Neural Bases of Human Language," *Yearbook of Physical Anthropology* 45 (2002): 36–62.

———, *Toward an Evolutionary Biology of Language* (Cambridge, Mass.: Belknap Press of Harvard University Press, 2006).

———, "The Evolution of Human Speech: Its Anatomical and Neural Bases," *Current Anthropology* 48 (2007): 39–66.

Lieberman, Philip, Angie Morey, Jesse Hochstadt, Marla Larson, and Sandra Mather, "Mount Everest: A Space-Analogue for Speech Monitoring of Cognitive Deficits and Stress," *Aviation, Space, and Environmental Medicine* 76 (2005): 198–207.

Locke, John L., and Barry Bogin, "Language and Life History: A New Perspective on the Development and Evolution of Human Language," *Behavioral and Brain Sciences* 29 (2006): 259–325.

Long, Kevin D., Grace Kennedy, and Evan Balaban, "Transferring an Inborn Auditory Perceptual Predisposition with Interspecies Brain Transplants," *PNAS* 98 (2001): 5862–67.

Longworth, C. E., S. E. Keenan, R. A. Barker, W. D. Marslen-Wilson, and L. K. Tyler, "The Basal Ganglia and Rule-Governed Language Use: Evidence from Vascular and Degenerative Conditions," *Brain* 128 (2005): 584–96.

Loritz, Donald, *How the Brain Evolved Language* (New York: Oxford University Press, 2002).

MacFarquhar, Larissa, "The Devil's Accountant," *The New Yorker* (March 21, 2003).

Maclay, Howard, "Overview," in eds. Danny D. Steinberg and Leon A. Jakobovits, *Semantics; an Interdisciplinary Reader in Philosophy, Linguistics and Psychology* (Cambridge, England: Cambridge University Press, 1971), p. 163.

MacNeilage, Peter F., "The Frame/Content Theory of Evolution of Speech Production," *Behavioral and Brain Sciences* 21 (1998): 499–546.

Maess, Burkhard, Stefan Koelsch, Thomas C. Gunter, and Angela D. Friederici, "Musical Syntax Is Processed in Broca's Area: An MEG Study," *Nature Neuroscience* 4 (2001): 540–45.

Marantz, Alec, Y. Miyashita, and Wayne O'Neil, *Image, Language, Brain: Papers from the First Mind Articulation Project Symposium* (Cambridge, Mass.: MIT Press, 2000).

Marcus, Gary F., "Before the Word," *Nature* 431 (2004): 745.

————, *The Birth of the Mind: How a Tiny Number of Genes Creates the Complexities of Human Thought* (New York: Basic Books, 2004).

————, "Startling Starlings," *Nature* 440 (2006): 1117–18.

Marcus, Gary F., and Simon E. Fisher, "Foxp2 in Focus: What Can Genes Tell Us About Speech and Language?" *Trends in Cognitive Sciences* 7 (2003): 257–62.

Marino, Lori, "Turning the Empirical Corner on Fi: The Probability of Complex Intelligence," in eds. Guillermo A. Lemarchand and Karen Jean Meech, *Bioastronomy '99: A New Era in Bioastronomy.* Proceedings of a meeting held at the Hapuna Beach Prince Hotel, Kohala Coast, Hawaii, 2–6 August, 1999 (San Francisco: Astronomical Society of the Pacific, 2000), pp. 431–35.

————, "Convergence of Complex Cognitive Abilities in Cetaceans and Primates," *Brain Behavior and Evolution* 59 (2002): 21–32.

————, "Absolute Brain Size: Did We Throw the Baby Out with the Bathwater?" *PNAS* 103 (2006): 13563–64.

Marslen-Wilson, William D., and Lorraine K. Tyler, "The Lexicon, Grammar, and the Past Tense: Dissociation Revisited," in eds. Michael Tomasello and Dan Isaac Slobin, *Beyond Nature-Nurture: Essays in Honor of Elizabeth Bates* (Mahwah, N.J.: L. Erlbaum, 2005).

Marx, J. L., "Ape-Language Controversy Flares Up," *Science* 207 (1980): 1330–33.

Matthews, P. H., *Grammatical Theory in the United States from Bloomfield to Chomsky* (Cambridge, England: Cambridge University Press, 1993).

Mayr, Ernst, *What Evolution Is* (New York: Basic Books, 2001).

McCowan, B., and D. Reiss, "The Fallacy of 'Signature Whistles' in Bottlenose Dolphins: A Comparative Perspective of 'Signature Information' in Animal Vocalizations," *Animal Behaviour* 62 (2001): 1151–62.

McNeill, D., B. Bertenthal, J. Cole, and S. Gallagher, "Gesture-First, but No Gestures?" *Behavioral and Brain Sciences* 28 (2005): 138–39.

McWhorter, John H., *The Power of Babel: A Natural History of Language* (New York: Times Books/Henry Holt, 2001).

Mekel-Bobrov, Nitzan, Sandra L. Gilbert, Patrick D. Evans, Eric J. Vallender, Jeffrey R. Anderson, Richard R. Hudson, Sarah A. Tishkoff, and Bruce T. Lahn, "Ongoing Adaptive Evolution of ASPM, a Brain Size Determinant in *Homo sapiens,*" *Science* 309 (2005): 1720–22.

Melis, Alicia P., Brian Hare, and Michael Tomasello, "Chimpanzees Recruit the Best Collaborators," *Science* 311 (2006): 1297–1300.

Mellars, Paul, "Why Did Modern Human Populations Disperse from Africa ca. 60,000 Years Ago? A New Model," *PNAS* 103 (2006): 9381–86.

Mercader, Julio, Huw Barton, Jason Gillespie, Jack Harris, Steven Kuhn, Robert Tyler, and Christophe Boesch. "4,300-Year-Old Chimpanzee Sites and the Origins of Percussive Stone Technology," *PNAS* 104 (2007): 3043–48.

Mercader, Julio, Melissa Panger, and Christophe Boesch, "Excavation of a Chimpanzee Stone Tool Site in the African Rainforest," *Science* 296 (2002): 1452–55.

Mietto, Paolo, Marco Avanzini, and Giuseppe Rolandi, "Human Footprints in Pleistocene Volcanic Ash," *Nature* 422 (2003): 133.

Morford, J. P,. "Insights to Language from the Study of Gesture: A Review of Research on the Gestural Communication of Non-Signing Deaf People," *Language and Communication* 16 (1996): 165–78.

Mulcahy, Nicholas J., and Josep Call, "Apes Save Tools for Future Use," *Science* 312 (2006): 1038–40.

Müller, Cornelia, "Gesture Studies: A New Field Emerging," paper presented at Gestural Communication in Nonhuman and Human Primates (workshop), Evolution of Language conference, Leipzig, Germany, March 2004.

Nadis, S., "Look Who's Talking," *New Scientist* (February 21, 2003).

Newport, Elissa L., Marc D. Hauser, Geertrui Spaepen, and Richard N. Aslin, "Learning at a Distance II. Statistical Learning of Non-Adjacent Dependencies in a Non-Human Primate," *Cognitive Psychology* 49 (2004): 85–117.

Nilsson, D.-E., and S. Pelger, "A Pessimistic Estimate of the Time Required for an Eye to Evolve," *Proceedings of the Royal Society of London B* 256 (1994): 53–58.

Nishimura, Takeshi, Akichika Mikami, Juri Suzuki, and Tetsuro Matsuzawa, "Descent of the Larynx in Chimpanzee Infants," *PNAS* 100 (2003): 6930–33.

Notman, Hugh, and Drew Rendall, "Contextual Variation in Chimpanzee Pant Hoots and Its Implications for Referential Communication," *Animal Behaviour* 70 (2005): 117–90.

Nowak, Martin A., Natalia L. Komarova, and Partha Niyogi, "Computational and Evolutionary Aspects of Language," *Nature* 417 (2002): 611–17.

O'Donnell, T. J., M. D. Hauser, and W. T. Fitch, "Using Mathematical Models of Language Experimentally," *Trends in Cognitive Sciences* 9 (2005): 284–89.

Oller, D. Kimbrough, and Ulrike Griebel, *Evolution of Communication Systems: A Comparative Approach* (Cambridge, Mass.: MIT Press, 2004).

Olsen, Steve, *Mapping Human History: Genes, Race and Our Common Origin* (New York: Houghton Mifflin, 2002).

Origgi, Gloria, and Dan Sperber, "A Pragmatic Perspective on the Evolution of Language and Languages," *Coevolution of Language and Theory of Mind* 6 (2004), http://www.interdisciplines.org/coevolution/papers/6.

Orr, H. Allen, "Darwinian Storytelling," *The New York Review of Books* 50 (February 27, 2003).

Oudeyer, Pierre-Yves, "The Non-Functional Origins of Phonetic Coding," paper presented at the From Animals to Animats 7: Proceedings of the Seventh International Conference on Simulation of Adaptive Behavior, Edinburgh, Scotland, August 2002.

———, "The Origins of Syllable Systems: An Operational Model," paper presented at the Proceedings of the 23rd International Conference on Cognitive Science, Edinburgh, Scotland, August 2001.

Özçalişkan, S., and S. Goldin-Meadow, "Gesture Is at the Cutting Edge of Early Language Development," *Cognition* 96 (2005): B101–B13.

Pääbo, Svante, "The Human Genome and Our View of Ourselves," *Science* 291 (2001): 1219–20.

————, "The Mosaic That Is Our Genome," *Nature* 421 (2003): 409–12.

Palumbi, Stephen R., "Humans as the World's Greatest Evolutionary Force," *Science* 293 (2001): 1786–90.

Patel, Aniruddh D., "Language, Music, Syntax and the Brain," *Nature Neuroscience* 6 (2003): 674–81.

Payne, Katharine, *Silent Thunder: In the Presence of Elephants* (New York: Simon & Schuster, 1998).

————, "The Progressively Changing Songs of Humpback Whales: A Window on the Creative Process in a Wild Animal," in eds. Björn Merker, Steven Brown, and Nils Lennart Wallin, *The Origins of Music* (Cambridge, Mass.: MIT Press, 2000), pp. 135–50.

————, "Sources of Social Complexity in the Three Elephant Species," in eds. Frans B. M. de Waal and Peter L. Tyack, *Animal Social Complexity: Intelligence, Culture, and Individualized Societies* (Cambridge, Mass.: Harvard University Press, 2003).

Pearce, Toby M., "Did They Talk Their Way Out of Africa?" *Behavioral and Brain Sciences* 26 (2003): 235–36.

Pennisi, Elizabeth, "Speaking in Tongues," *Science* 303 (2004): 1321–23.

Pepperberg, I. M., and H. R. Shive, "Simultaneous Development of Vocal and Physical Object Combinations by a Grey Parrot (*Psittacus erithacus*): Bottle Caps, Lids, and Labels," *Journal of Comparative Psychology* 115 (2001): 376–84.

Pepperberg, I. M., and S. E. Wilcox, "Evidence for a Form of Mutual Exclusivity During Label Acquisition by Grey Parrots (*Psittacus erithacus*)?" *Journal of Comparative Psychology* 114 (2000): 219–31.

Peretz, Isabelle, and Max Coltheart, "Modularity of Music Processing," *Nature Neuroscience* 6 (2003): 688–91.

Phillips-Silver, Jessica, and Laurel J. Trainor, "Feeling the Beat: Movement Influences Infant Rhythm Perception," *Science* 308 (2005): 1430.

Piattelli-Palmarini, Massimo, *Language and Learning: The Debate Between Jean Piaget and Noam Chomsky* (Cambridge, Mass.: Harvard University Press, 1980).

Pinker, Steven, *The Language Instinct* (New York: William Morrow, 1994).

————, *How the Mind Works* (New York: W. W. Norton, 1999).

————, "Talk of Genetics and Vice Versa," *Nature* 413 (2001): 465–66.

————, *The Blank Slate: The Modern Denial of Human Nature* (London: Allen Lane, 2002).

————, " 'Words and Rules': An Exchange," *The New York Review of Books* 49 (June 27, 2002).

————, " 'The Blank Slate': An Exchange," *The New York Review of Books* 50 (May 1, 2003).

Pinker, Steven, and Paul Bloom, "Natural Language and Natural Selection," *Behavioral and Brain Sciences* 13 (1990): 707–84.

Pinker, Steven, and Ray Jackendoff, "The Faculty of Language: What's Special About It?" *Cognition* 95 (2005): 201–36.

Plotnik, Joshua M., Frans B. M. de Waal, and Diana Reiss, "Self-Recognition in an Asian Elephant," *PNAS* 103 (2006): 17053–57.

Poole, Joyce H., Katherine Payne, William R. Langbauer, Jr., and Cynthia J. Moss, "The Social Context of Some Very Low Frequency Calls of African Elephants," *Behavorial Ecology and Sociobiology* 22 (1988): 385–92.

Poole, Joyce H., Peter L. Tyack, Angela S. Stoeger-Horwath, and Stephanie Watwood, "Animal Behaviour: Elephants Are Capable of Vocal Learning," *Nature* 434 (2005): 455–56.

Poremba, Amy, Megan Malloy, Richard C. Saunders, Richard E. Carson, Peter Herscovitch, and Mortimer Mishkin, "Species-Specific Calls Evoke Asymmetric Activity in the Monkey's Temporal Poles," *Nature* 427 (2004): 448–51.

Premack, David, "Is Language the Key to Human Intelligence?" *Science* 303 (2004): 318–20.

Pruetz, Jill D., and Paco Bertolani, "Savanna Chimpanzees, Pan Troglodytes Verus, Hunt with Tools," *Current Biology* 17 (2007): 1–6.

Ramus, Franck, Marc D. Hauser, Cory Miller, Dylan Morris, and Jacques Mehler, "Language Discrimination by Human Newborns and by Cotton-Top Tamarin Monkeys," *Science* 288 (2000): 349–51.

Reiss, Diana, and Lori Marino, "Mirror Self-Recognition in the Bottlenose Dolphin: A Case of Cognitive Convergence," *PNAS* 98 (2001): 5937–42.

Renfrew, Colin, "At the Edge of Knowability: Towards a Prehistory of Languages," *Cambridge Archaeological Journal* 10 (2000): 7–34.

Roeper, Thomas, "The Chomsky Experiments," *The New York Review of Books* 28 (April 16, 1981).

Rousseau, Jean-Jacques, and Victor Gourevitch, *The First and Second Discourses Together with the Replies to Critics and Essay on the Origin of Languages* (New York: HarperCollins, 1990).

Rumbaugh, D. M., W. D. Hopkins, D. A. Washburn, and E. S. Savage-Rumbaugh, "Comparative Perspectives of Brain, Cognition, and Language," in ed. Norman A. Krasnegor, *Biological and Behavioral Determinants of Language Development* (Hillsdale, N.J.: L. Erlbaum, 1991), pp. 145–46.

Salzberg, Steven L., Owen White, Jeremy Peterson, and Jonathan A. Eisen, "Microbial Genes in the Human Genome: Lateral Transfer or Gene Loss?" *Science* 292 (2001): 1903–6.

Sanders, Ira, "Human Tongue, Pharynx and Vocal Fold Muscles Contain Slow Tonic Muscle," paper presented at Evolution of Language Conference, Harvard University, Cambridge, Mass., March 27, 2002.

Sandler, Wendy, Irit Meir, Carol Padden, and Mark Aronoff, "The Emergence of Grammar: Systematic Structure in a New Language," *PNAS* 102 (2005): 2661–65.

Sapolsky, Robert M., *A Primate's Memoir* (New York: Scribner, 2001).

Savage-Rumbaugh, E. S., "Why Are We Afraid of Apes with Language?" in eds. Arnold B. Scheibel and J. William Schopf, *The Origin and Evolution of Intelligence* (Boston: Jones and Bartlett, 1997), pp. 43–69.

Savage-Rumbaugh, E. Sue, and Roger Lewin, *Kanzi: The Ape at the Brink of the Human Mind* (New York: Wiley, 1994).

Savage-Rumbaugh, E. S., and D. M. Rumbaugh, "Perspectives on Consciousness, Language, and Other Emergent Processes in Apes and Humans," in eds. Stuart R. Hameroff, Alfred W. Kaszniak, and Alwyn Scott, *Toward a Science of Consciousness II: The Second Tucson Discussions and Debates* (Cambridge, Mass.: MIT Press, 1998), pp. 533–49.

Savage-Rumbaugh, S., S. Shanker, and T. J. Taylor, "Apes with Language," *Critical Quarterly* 38 (1996): 45–57.

Savage-Rumbaugh, E. Sue, Stuart Shanker, and Talbot J. Taylor, *Apes, Language, and the Human Mind* (New York: Oxford University Press, 1998).

Schusterman, Ronald J., Colleen Reichmuth Kastak, and David Kastak, "The Cognitive Sea Lion: Meaning and Memory in the Laboratory and in Nature," in eds. Marc Bekoff, Colin Allen, and Gordon M. Burghardt, *The Cognitive Animal: Empirical and Theoretical Perspectives on Animal Cognition* (Cambridge, Mass.: MIT Press, 2002).

Schwartz, D. A., C. Q. Howe, and D. Purves, "The Statistical Structure of Human Speech Sounds Predicts Musical Universals," *Journal of Neuroscience* 23 (2003): 7160–68.

Searle, John R., "A Special Supplement: Chomsky's Revolution in Linguistics," *The New York Review of Books* 18 (June 29, 1972).

———, "End of the Revolution," *The New York Review of Books* 49 (February 28, 2002).

Searls, David B., "Trees of Life and of Language," *Nature* 426 (2003): 391–92.

Seyfarth, Robert M., and Dorothy L. Cheney, "Signallers and Receivers in Animal Communication," *Annual Review of Psychology* 54 (2003): 145–73.

Seyfarth, Robert M., Dorothy L. Cheney, and Thore J. Bergman, "Primate Social Cognition and the Origins of Language," *Trends in Cognitive Sciences* 9 (2005): 264–66.

Seyfarth, R. M., D. L. Cheney, and P. Marler, "Monkey Responses to Three Different Alarm Calls: Evidence of Predator Classification and Semantic Communication," *Science* 210 (1980): 801–3.

Shapiro, Kevin A., Lauren R. Moo, and Alfonso Caramazza, "Cortical Signatures of Noun and Verb Production," *PNAS* 103 (2006): 1644–49.

Shu, Weiguo, Julie Y. Cho, Yuhui Jiang, Minhua Zhang, Donald Weisz, Gregory A. Elder, James Schmeidler, Rita De Gasperi, Miguel A. Gama Sosa, Donald Rabidou, Anthony C. Santucci, Daniel Perl, Edward Morrisey, and Joseph D. Buxbaum, "Altered Ultrasonic Vocalization in Mice with a Disruption in the Foxp2 Gene," *PNAS* 102 (2005): 9643–48.

Silk, Joan B., "Who Are More Helpful, Humans or Chimpanzees?" *Science* 311 (2006): 1248–49.

Siveter, David J., Mark D. Sutton, Derek E. G. Briggs, and Derek J. Siveter, "An Ostracode Crustacean with Soft Parts from the Lower Silurian," *Science* 302 (2003): 1749–51.

Skinner, B. F., *Verbal Behavior* (New York,: Appleton-Century-Crofts, 1957).

———, *Beyond Freedom and Dignity* (New York: Bantam/Vintage, 1971).

Steels, Luc, *The Talking Heads Experiment*, vol. 1: *Words and Meanings* (Antwerp, Belgium: Laboratorium, 1999).

———, "Language as a Complex Adaptive System," *Lecture Notes in Computer Science* (2000): 17–28.

———, "Mirror Neurons and the Action Theory of Language Origins," paper presented at Architectures of the Mind, Architectures of the Brain Conference, September 2000, http://www.csl.sony.fr/General/Publications/ByAuthor.php?author=steels&year =2000.

———, "The Recruitment Theory of Language Origins," paper presented at Morris Symposium on Language Evolution, Stony Brook, N.Y., October 2005.

Steels, L., and F. Kaplan, "Aibo's First Words: The Social Learning of Language," *Evolution of Communication* 4 (2001): 3–32.

Steels, L., F. Kaplan, A. McIntyre, and J. Van Looveren, "Crucial Factors in the Origins of Word-Meaning," in ed. Alison Wray, *The Transition to Language* (Oxford: Oxford University Press, 2002), pp. 252–71.

Stefansson, Hreinn, Agnar Helgason, Gudmar Thorleifsson, Valgerdur Steinthorsdottir, Gisli Masson, John Barnard, Adam Baker, Aslaug Jonasdottir, Andres Ingason, Vala G. Gudnadottir, Natasa Desnica, Andrew Hicks, Arnaldur Gylfason, Daniel F. Gudbjartsson, Gudrun M. Jonsdottir, Jesus Sainz, Kari Agnarsson, Birgitta Birgisdottir, Shyamali Ghosh, Adalheidur Olafsdottir, Jean-Baptiste Cazier, Kristleifur Kristjansson, Michael L. Frigge, Thorgeir E. Thorgeirsson, Jeffrey R. Gulcher, Augustine Kong, and Kari Stefansso, "A Common Inversion Under Selection in Europeans," *Nature Genetics* 37 (2005): 129–37.

Studdert-Kennedy, Michael, "Vocal Imitation, Facial Imitation, and the Gestural Origin of Linguistic Discrete Infinity," paper presented at Evolution of Language conference, Harvard University, Cambridge, Mass. March 27, 2002.

Sutherland, William J., "Parallel Extinction Risk and Global Distribution of Languages and Species," *Nature* 423 (2003): 276–79.

Tattersall, Ian, *The Monkey in the Mirror: Essays on the Science of What Makes Us Human* (New York: Harcourt, 2002).

———, "Once We Were Not Alone," *Scientific American*, May 1, 2003.

Teotónio, Henrique, and Michael R. Rose, "Variation in the Reversibility of Evolution," *Nature* 408 (2000): 463–66.

Terrace, H. S., L. A. Petitto, R. J. Sanders, and T. G. Bever, "Can an Ape Create a Sentence?" *Science* 206 (1979): 891–902.

Tincoff, Ruth, and Marc D. Hauser, "Cognitive Basis for Language Evolution in Nonhuman Primates," in eds. E. K. Brown and Anne Anderson, *Encyclopedia of Language and Linguistics* (Amsterdam and Boston: Elsevier, 2005).

Tincoff, Ruth, Marc Hauser, Fritz Tsao, Geertrui Spaepen, Franck Ramus, and Jacques Mehler, "The Role of Speech Rhythm in Language Discrimination: Further Tests with a Non-Human Primate," *Developmental Science* 8 (2005): 26–35.

Tolbert, Elizabeth, "Music and Meaning: An Evolutionary Story," *Psychology of Music* 29 (2001): 84–94.

Tomasello, Michael, "Why Don't Apes Point?" in eds. N. Endfield and S. Levinson, *Roots of Human Sociality* (New York: Werner-Gren, in press).

Tomasello, M., M. Carpenter, J. Call, T. Behne, and H. Moll, "Understanding and Sharing Intentions: The Origin of Cultural Cognition," *Behavioral and Brain Sciences* 28 (2005): 675–90.

Trehub, Sandra E., "The Developmental Origins of Musicality," *Nature Neuroscience* 6 (2003): 669–73.

Tremblay, Stephanie, Douglas M. Shiller, and David J. Ostry, "Somatosensory Basis of Speech Production," *Nature* 423 (2003): 866–69.

Tyler, L. K., W. D. Marslen-Wilson, and E. A. Stamatakis, "Differentiating Lexical Form, Meaning, and Structure in the Neural Language System," *PNAS* 102 (2005): 8375–80.

Umiker-Sebeok, Jean, and Thomas A. Sebeok, "More on Monkey Talk," *The New York Review of Books* 27 (December 4, 1980).

Underhill, Peter A., Peidong Shen, Alice A. Lin, Li Jin, Giuseppe Passarino, Wei H. Yang, Erin Kauffman, Batsheva Bonne-Tamir, Jaume Bertranpetit, Paolo Francalacci, Muntaser Ibrahim, Trefor Jenkins, Judith R. Kidd, S. Qasim Mehdi, Mark T. Seielstad, R. Spencer Wells, Alberto Piazza, Ronald W. Davis, Marcus W. Feldman, L. Luca Cavalli-Sforza, and Peter J. Oefner, "Y Chromosome Sequence Variation and the History of Human Populations," *Nature Genetics* 26 (2000): 358–61.

van Schaik, Carel, "Why Are Some Animals So Smart?" *Scientific American* (March 26, 2006).

Vargha-Khadem, F., L. J. Carr, E. Isaacs, E. Brett, C. Adams, and M. Mishkin, "Onset of Speech After Left Hemispherectomy in a Nine-Year-Old Boy," *Brain* 120 (1997): 159–82.

Vargha-Khadem, F., K. Watkins, K. Alcock, P. Fletcher, and R. Passingham, "Praxic and Nonverbal Cognitive Deficits in a Large Family with a Genetically Transmitted Speech and Language Disorder," *PNAS* 92 (1995): 930–33.

Vargha-Khadem, F., K. E. Watkins, C. J. Price, J. Ashburner, K. J. Alcock, A. Connelly, R. S. J. Frackowiak, K. J. Friston, M. E. Pembrey, M. Mishkin, D. G. Gadian, and R. E. Passingham, "Neural Basis of an Inherited Speech and Language Disorder," *PNAS* 95 (1998): 12695–700.

Varley, Rosemary A., Nicolai J. C. Klessinger, Charles A. J. Romanowski, and Michael Siegal, "Agrammatic but Numerate," *PNAS* 102 (2005): 3519–24.

Visalberghi, E., D. M. Fragaszy, and S. Savage-Rumbaugh, "Performance in a Tool-Using Task by Common Chimpanzees (*Pan troglodytes*), Bonobos (*Pan paniscus*), an Orangutan (*Pongo pygmaeus*) and Capuchin Monkeys (*Cebus apella*)," *Journal of Comparative Psychology* 109 (1995): 52–60.

Volterra, Virginia, Maria Cristina Caselli, Olga Capirci, and Elena Pizzuto, "Gesture and the Emergence and Development of Language," in eds. Michael Tomasello and Dan Isaac Slobin, *Beyond Nature-Nurture: Essays in Honor of Elizabeth Bates* (Mahwah, N.J.: Lawrence Erlbaum Associates Publishers, 2005).

Wade, Nicholas, "Still Evolving, Human Genes Tell New Story," *The New York Times*, March 7, 2006.

Warneken, Felix, and Michael Tomasello, "Altruistic Helping in Human Infants and Young Chimpanzees," *Science* 311 (2006): 1301–3.

Wells, Spencer, *The Journey of Man: A Genetic Odyssey* (London: Penguin, 2003).

Wen, Q., and D. B. Chklovskii, "Segregation of the Brain into Gray and White Matter: A Design Minimizing Conduction Delays," *PLoS Computational Biology* 1 (2005): e78.

Whiten, A., J. Goodall, W. C. McGrew, T. Nishida, V. Reynolds, Y. Sugiyama, C. E. G. Tutin, R. W. Wrangham, and C. Boesch, "Cultures in Chimpanzees," *Nature* 399 (1999): 682–85.

Whiting, Michael F., Sven Bradler, and Taylor Maxwell, "Loss and Recovery of Wings in Stick Insects," *Nature* 421 (2003): 264–67.

Williams, Bernard, "Where Chomsky Stands," *The New York Review of Books* 23 (November 11, 1976).

Williams, S. L., K. E. Brakke, and E. S. Savage-Rumbaugh, "Comprehension Skills of

Bibliography

Language-Competent and Nonlanguage-Competent Apes," *Language & Communication* 17 (1997): 301–17.

Wolpoff, Milford H., Brigitte Senut, Martin Pickford, and John Hawks, "Sahelanthropus or 'Sahelpithecus'?" *Nature* 419 (2002): 581–82.

Wong, Kate, "Who Were the Neandertals?" *Scientific American*, April 10, 2000.

Wrangham, Richard W., *Chimpanzee Cultures* (Cambridge, Mass.: Harvard University Press in cooperation with the Chicago Academy of Sciences, 1994).

Wynn, Karen, "Addition and Subtraction by Human Infants," *Nature* 358 (1992): 749–50.

Yang, Charles, "Dig-Dig, Think-Thunk," *London Review of Books*, 24 August 2000.

Zuberbühler, Klaus, "A Syntactic Rule in Forest Monkey Communication," *Animal Behaviour* 62 (2002): 293–99.

Zuckerman, Lord, "Apes Я Not Us," *The New York Review of Books* 38 (May 30, 1991).

INDEX